To Andy,
A great member of a great family!

A Silent Warrior Steps Out of the Shadows

Memoir of CDR Guy Thomas, USN (ret.)

Intelligence Operator and Innovator

From Deck Force Seaman to Space Operations

Guy Thomas

Copyright © 2021 by George Guy Thomas

All rights reserved. No part of this publication may be reproduced, distributed, or transmitted in any form or by any means, including photocopying, recording, or other electronic or mechanical methods, without the prior written permission of the publisher, except in the case of brief quotations embodied in critical reviews and certain other noncommercial uses permitted by copyright law. For permission requests, write to the publisher, addressed "Attention: Permissions Coordinator," at the address below.

Alpha Book Publisher
www.alphapublisher.com

Ordering Information:
Quantity sales. Special discounts are available on quantity purchases by corporations, associations, and others.
For details, contact the publisher at the address above.
Orders by U.S. trade bookstores and wholesalers.
Visit www.alphapublisher.com/contact-us to learn more.

Printed in the United States of America

Table of Contents

Introduction: Setting the Stage 1
Chapter 1: Looking Back from the Mid-Point 7
Chapter 2: Early Times, then in the Navy 14
Chapter 3: USS HORNE – A Missile Cruiser 25
Chapter 4: Shrike Hit! - My First Watch 51
Chapter 5: Mines, Missiles, and MiG's 71
Chapter 6: Airborne Reconnaissance 94
Chapter 7: Submarines at last! 113
Chapter 8: Experience Pays Off! 133
Chapter 9: My First (and Only) Command 145
Chapter 10: Career Takes off with the USAF 164
Chapter 11: Naval War College – Student, Innovator 204
Chapter 12: Joint Electronic Warfare Center 227
Chapter 13: A Dog's Breakfast 242
Chapter 14: FBI, Undercover - Short but Sweet! ... 258
Chapter 15: Origins of Satellite AIS - My Child! .. 271
Annex 1 301
Annex 2 333
Annex 3 336

A Silent Warrior
Steps
Out of the Shadows

CDR Geo. Guy Thomas, USN (ret.)

Setting the Stage

The title of this memoir comes from two related places. First, "back in the day," those of us who served in military reconnaissance (AKA intelligence collection, AKA Spying) called ourselves "Silent Warriors" because we could tell no one, not even our families, what we were doing in any detail. Our families knew we were engaged in dangerous operations, but we could never discuss specifics at all. Thus, the stress on the families was, and probably still is, immense. It is one of the significant sacrifices "Silent Warriors" make. Second, members of the intelligence world operate "in the shadows."

It is hard for me to believe my career happened, and even harder, that I lived through it! I started as a deck-force seaman on a missile cruiser. After a successful career in the intelligence world and then in the industry, I ended up as an honored member of both the classified and unclassified space communities. The twisted and convoluted road between those two points is the story I am going to relate here.

Toward the end of my career, while an employee of Johns Hopkins University's Applied Physics Lab (JHU/APL), I conceived satellite AIS (S-AIS). I have spent the last 19 years working on first getting it funded and built and then making the world aware of the unique tool it has become. But the first 23 years of my adult career were very different. I spent those years in an almost unknown part of the Navy, where I was deeply involved in highly classified and sometimes dangerous operations.

I thought I would die by being shot down, crash, or sunk over a dozen times during my ten years of field operations. Almost 11, if you count my time with the FBI as a drafted volunteer. A vivid imagination is not a good thing to have in some situations, but, "if you are not at least a little bit scared, I do not want to fly a mission with you," is an adage in military aviation. That definitely pertains to the reconnaissance business. If you are a bit scared, you are focused. If you are not focused, you can get killed in a hurry. You do learn to operate a bit concerned. That is good. It keeps you alert.

But that is part of my early history. Let us return to the present. S-AIS is now also used for a great deal more than I had initially foreseen. The same thing happened to satellite navigation, the forerunner of GPS. I was an employee of JHU/APL when I conceived S-AIS, and JHU/APL is also where satellite navigation was created. I believe this is not a coincidence.

JHU/APL is a unique place. I spent nine years there. The intellectual ferment at APL is unlike any other organization I know. The Naval War College's research side, the Center for Naval Warfare Studies (CNWS), where I also spent over seven years, comes close. Still, APL is exceptional. It adds the science and technology development, making the ideas generated at NWC and CNWS implementable. It was an exciting place to work, and I seemed to crave excitement.

It is fair to ask why I think that. And also, to ask, how did I, a former deck-force seaman and college student studying history and hoping to go to law school someday, come to be at JHU/APL, the Navy's center of technology innovation and creation, in the first place.

My course from the deck force of a missile cruiser to APL was anything but a straight line. I was a Navy "spook" for 20 exciting years. "Spook" is a nickname for someone who collects intelligence, although some also use the name to refer to those who analyze intelligence as well. I was a genuine spook. I went in harm's way to collect intelligence. When I became too senior for work in the field, I worked on improving one aspect or another of the United States' ability to collect, fuse, analyze or report intelligence for most of my 23-year military career.

I was a member of the elite Naval Security Group (NavSecGru). I say elite because many people told me that multiple times, in many places. I know that the NavSecGru had the first pick of enlisted personnel and the second pick of newly minted officers, just behind the nuclear submarine program. Its enlisted personnel had to score higher on the Navy skills aptitude and intelligence tests than the score to get into the Naval Academy.

That raises the question as to how I got to be a Spook. That question is, in many ways, the heart of this story. I was put on this course by a series of events that I initially considered bad luck. I think I probably squeezed in because of my experience on that missile cruiser off Vietnam (part of the "bad Luck") as an enlisted man (more "bad luck"), where I became the acting Intelligence Officer (balancing good luck!). And I do take aptitude tests very well.

Since World War II, the Naval Security Group has had teams of intercept operators, analysts, and sometimes communicators deployed on ships, submarines, and aircraft to provide signals intelligence (SIGINT) in direct support (DirSup) of operational units in hostile areas. Because these units were forward deployed and engaged in activities that often drew hostile reactions, they were dangerous. At times people died, but they were also unique collection opportunities. These units were (and probably still are) front-line intelligence collection units.

To the best of my knowledge, and that of Admiral Bobbie Inman, who I conferred with while writing this story, I am the first career officer who served in these tactical/operational teams targeted against hostile forces to ever write about these operations in any detail. They were highly classified during my time in them.

Indeed, the very existence of these teams was classified 40+ years ago. However, times have changed. In the last few years, I have seen SIGINT collection mentioned in the open press, including in press interviews with very high-ranking officers in the intelligence community. So, I decided to "test the water" and see if I could get my story, rather unique even for a "Spook," approved by the proper authorities. Because my time in these operations was long enough ago, I suspected I could now talk about my time in those teams that collected that intelligence without jeopardizing anyone or anything today.

We were highly trained to do our job, especially if you came through the submarine pipeline, as I did. Even in the military and fewer yet in the civilian world few people understand how we trained, what we did, where our career path might take us, or what career highlights might entail.

This book is an effort to pull back the cloak of secrecy just a bit and describe these things from a personal view without violating national security or any legal constraints. My military career was one of the most exciting as well as the oddest I know. I didn't plan it that way; it just happened, often with me thinking I was being screwed. But, in looking back, nearly all these trying twists of my career path (AKA "screwings") were actually "blessings in disguise."

I participated in reconnaissance and surveillance operations for over ten years, something like three times the average of most NavSecGru officers or officers of any other service for that matter. However, the previously enlisted officers are an exceptional group, and I take my hat off to them. Indeed, I am proud I helped a number of them make the enlisted to officer transition. In many ways, it was the highlight of my life. It is undoubtedly a fact that, except for my creation of S-AIS, as I mentioned above, it is my time in the Navy and especially

NavSecGru, as unusual as it was, of which I am most proud.

My career since I retired from the Navy has been odder yet. I have had several other adventures since I retired from the Navy, but the two peaks of my post-Navy career were dramatically different.

The first came five years after I retired from the Navy. I lived in the Dallas area, where I spent several months working for the FBI in an undercover role, observing some suspicious individuals I had brought to their attention. Nine years later, the FBI arrested more than 20 spies of Hezbollah and Hamas due, in part, to my early work.

Of course, as I mentioned above, the absolute highpoint of my life came on 4 October 2001 when, at 58, I conceived satellite AIS (S-AIS) in a flash of intuition. That intuition, based on 35+ years of experience and study, traces back to my training and experience as a Navy "Spook" as well as my time with the US Air Force.

I have included why and how I created satellite AIS (S-AIS) because my time working on S-AIS also involved me in historical events in the civilian maritime world. It took me almost three years to find funding for S-AIS and get it on track for full development and deployment, and there was a lot of frustration in those years, but the payoff has been HUGE.

My idea to take an Automatic Identification System (AIS) receiver, designed as collision avoidance and local maritime traffic management beacon between ships, and put it on a satellite to be used as a global ship tracking system was initially met with great skepticism and even good-natured ridicule. However, I was sure I was right, at least in part because of my odd set of experiences, so I persevered. I am very glad I did.

There are now over 205 AIS receivers in Space and more on the way. S-AIS now generates accurate updates of the location of nearly every *legal* ship anywhere in the world, almost once a minute. Many people involved with civilian maritime operations believe S-AIS has caused the most significant paradigm shift in the world's maritime operations since the steam engine, the screw propeller, or radar. Yes, even more important than satellite navigational even in its latest form, GPS.

A Royal Navy commodore recently compared the invention of S-AIS to John Harrison's creation of the chronometer, which allowed mariners to calculate their longitude. Mariners had known their latitude by the elevation of the North Star above the horizon for hundreds of years. Still, until John Harrison's chronometer, they had no way of precisely knowing "local apparent noon," which was crucial to determining longitude, until satellite navigation. More on the impact of S-AIS later in this book.

I am also in the process of writing a second book I have tentatively entitled "Satellite AIS and the Birth of Global Maritime Awareness, How and Why they both came to be." It is the story of my life since 9/11, since I left "the Shadows." I am utterly amazed by my second career's uniqueness as well. I am a blessed man for some unknowable reason. ¡Gracias de Dios!

The "FLISH"
Emblem of the Naval Security Group
Direct Support Teams

A Mid-Point after Early Commitment: June 1978

Chapter One

As I came down the ladder leaving the EP-3E reconnaissance aircraft (AKA Spy plane) there at Naval Air Facility Atsugi, Japan, I felt a bit sad and nostalgic and proud.

I knew this was probably going to be my last operational flight in the Navy. It was late June 1978, and I was completing my 27-month tour as Officer-in-Charge, Naval Security Group Detachment (NavSecGruDet) Atsugi. I had been in Japan for almost seven years, and it had been quite a run, both professionally and personally.

As I reached the bottom rung of the ladder on that scorching and sticky day, I suddenly had a flashback to another sweltering and humid day nearly 20 years earlier. That day in late July 1958 had changed my life even though nothing momentous happened then. Still, there on the airfield in Japan, I suddenly remembered that earlier, also a blazingly hot day in Texas, with crystal clarity.

I was returning from visiting the tiny grocery store on Lawnview between Indiana and Ohio Streets in Corpus Christi, Texas, about four blocks from our home at 338 Atlantic. As I approached Ohio Street that day, walking home licking my ice cream, I suddenly realized I was really, really bored. Not an unusual feeling for a nearly 15-year-old boy, but what I did to relieve it, I now realize many years later, was.

In May of that year, I became an Eagle Scout, something I am still immensely proud of all these many years later. It had taken nearly three years of focused effort. I worked on some aspects of the many requirements almost every day, especially during the last 20 months when I changed to Troop 216 under a great scoutmaster, Mr. Herb Noakes.

In June, I had gone to the regular scout camp as a junior leader for twelve days and then to a wilderness camp in the West Texas mountains, something I had thoroughly enjoyed. In early July, I had gone to Camp Capers, an Episcopal Church camp in the beautiful Texas Hill Country, where my team had won the "Square C," the award for being best "cabin" (crew/team), at least in part due to my Boy Scout leadership training and experience. It had been a great summer until mid-July but now, after two weeks with no goal or anything specific to focus my energy on, I was very bored.

EP-3E Aeries II

As I walked down Lawnview, a dusty little side street with crushed shell sidewalks, I realized I needed a new goal to focus on, but what? As I pondered what to do next, I recalled listening to my older brother, Bob. He was 19 at the time and considered to be "the brains of the family." He was a sophomore at the prestigious University of the South, Sewanee, Tennessee, via a generous complete scholarship. My family certainly did not have the money to pay for school tuition such as that.

But it was not school, girls, sports, or cars he and his friends discussed every day, but rather how they would fulfill their military requirement. It was their primary subject of discussion.

In those days, all 18-year-old "able-bodied males" had a 2-year obligation to serve their nation. (This is still a good idea to me, except I would include females in the Draft as well and offer service in such things as the Peace Corps and Volunteers in Service to America as alternative ways to serve our great nation.) If the 18-year-old "men" were not a full-time college student, they were eligible for "the DRAFT."

They could wait to be drafted, or visit the various recruiters; take a battery of aptitude tests, and, determining how well they placed, agree to go in the military for more years than the two years the Draft required, in exchange for technical training: jet engine mechanic, nurse, heavy equipment operator or maintenance, pipefitter, welder, electrician, policeman, etc. Of course, each service had different needs, and there was much discussion among all young men in those days as to what was the best deal and the best service.

With those discussions in mind and influenced, no doubt, by my substantial reading on naval and maritime subjects, my favorite reading material by a long way in those days, I decided then and there on Lawnview, that I was going to go in the Navy, probably for a career, if I could measure up.

An immediate second question popped up. What did I want to do in the Navy? I had read "Navy Wings of Gold" and other stories of the carrier war in the Pacific and immediately thought of being a naval aviator. However, those men were made out to be supermen and I just might not measure up, so what were my other alternatives?

I had read "Run Silent, Run Deep" and other books about submarine warfare in World War II. I distinctly remembered many parts of those stories as I walked down Lawnview. I thought it would be very cool to be a submariner, especially in a nuclear submarine. However, the nuclear submarines' crews were the nearest thing we had to national heroes in those pre-NASA days before astronauts.

Again, I felt I might not measure up, so I thought a bit more. I recalled reading "The Boy's Book of Sea Battles" (the first book I ever read on my own!) and C. S. Forester and Alister McLean's books on surface warfare, both in wooden and iron ships. I thought it would be grand to serve in destroyers and cruisers.

Finally, in many of the books I read on naval warfare, the role of naval intelligence officers was crucial. So, I also added that role to my list for consideration. You might as well aim high, as my Father always said! Lastly, tales of living overseas and the value of learning a foreign language and culture were also interspersed in many of those stories, so I added: "Live overseas for a significant amount of time and learn a foreign language" to my shortlist.

Most of my six uncles had served in World War II, which had ended less than 15 years earlier. I had heard many stories from them, but I still did not know much about the military profession, especially about careers in the Navy at that moment. I suspected that I was looking at four mutually exclusive sets: Aviator, Submariner, Surface officer, Intel. I guessed that I might be able to combine one of the warfare specialties with being an intelligence officer, but that would be about the extent of it. So, I decided then and there on Lawnview that I would join the Sea Scouts to learn the basics of seamanship and the maritime world. And I put the final decision as to what my career in the Navy would be in God's hands. I am not an overly religious person, but I do acknowledge a supreme being and what happened in the next 30 years is a clear testament to me that there is one, and He (She?) has smiled on me for some unknowable reason.

As I stepped off the ladder of that EP-3E spy plane for the last time, the newest airplane in the Navy's inventory, I was stopped dead in my tracks. The EP-3E was the first aircraft in the world to use a central computer to run a primary mission system, and I had been responsible for wringing out that system and bringing it into service. It was a big task, but that was not what had stopped me in my tracks. My mind had flashed back, unbidden, to that hot late July day in Texas nearly 20 years earlier.

"Holy S----! I have done everything I dreamed of nearly 20 years ago." I had just finished nearly seven years forward deployed to Japan, operating on the ragged edge of the Cold War. In the reconnaissance and intelligence collection business, the Cold War was always just a short step away from going hot, and we all knew it.

Over 200 men in my specific part of the Navy, signals intelligence and reconnaissance, had been killed or captured since I joined the Navy doing what I had just finished doing for more years than any other officer I knew. Two ships, USS Pueblo and USS Liberty, and one aircraft, an EC-121 from my squadron, Fleet Air Reconnaissance One (VQ-1), had been captured, severely wounded, or shot down respectively since I joined the Navy.

But that was not what gave me pause. It was the fact that I had served in three missile cruisers, USS Horne (DLG-30), USS Worden (DLG-18), and USS Chicago (CG-11) in combat, and two submarines, USS Pintado (SSN-672) and USS Pogy (SSN-647), on classified missions that could have, and very nearly did on several occasions, turn into actual combat.

Additionally, I had flown on three Navy reconnaissance aircraft, the EC-121M, EP-3B, and EP-3E, on dangerous missions of national importance in hostile airspace, well within the range of potential enemy weapons systems. I also had verbal orders to USS Halibut (SSN- 589) and then her replacement, USS Seawolf (SSN-575), as the intelligence collection team chief on a very special-special (sic) mission. My orders to go to sea on that very special-special mission were canceled at near the last minute when it became necessary to send someone with at least some experience with computers to Atsugi on short notice to oversee the acceptance testing of the EP-3E.

Since joining the Navy, I had taken five computer science courses, two as an enlisted man and three since becoming an officer. Thus, I was the only one, both semi-qualified and semi-available, as well as cleared and flight qualified. Finally, I was now a designated signals intelligence officer certified in several systems. I also spoke reasonable Russian and Japanese. I HAD DONE EVERYTHING I CONCEIVED OF DOING 19 years and 11 months earlier! Thank you, GOD! Then I said something that, in looking back, was very foolish, but I did not know it at the time. "God, realize the rest of my life is going to be pretty tame, but I acknowledge that I owe you a Big One, and I will not complain. Thank you very, very much!"

I believed at that moment and still do that I was possibly the luckiest man alive. My life previous to that moment had not been a complete bed of roses, but, on balance, "the prize was worth the game." Just over a month after that flashback on the airfield at Atsugi, I reported to the US Air Force Security Service Headquarters, Kelly Air Force Base, San Antonio, Texas as the Navy exchange officer. It was a plumb assignment, and my professional life took off in many ways, no pun intended.

Less than two months after I reported on board Kelly AFB, I almost crashed in a US Air Force RC-135M. We were well out in international airspace over the northern Gulf of Tonkin when four late-model MiG- 21Hs launched to intercept us. We almost crashed into the sea diving to evade them, leveling out at 200 feet rather than the planned 500 feet. My anticipated "boring life" has been put on hold, but I am getting way ahead of myself. Let me just say that the "boring life" I was anticipating has NEVER materialized. Let us go back and describe my Navy career and life up to that point before we go forward.

Early Times then off to Boot Camp

Chapter Two

Informal preparation, fueled by curiosity, has always been a hallmark of mine. I now realize that much of the "Good Luck" I have had was because I did have that informal preparation based on intellectual curiosity. Clearly, that axiom attributed to storied Alabama Football coach Bear Bryant that goes: "Have you ever noticed that it is the well-prepared teams that seem to have all the luck?" applies to me in spades.

**This Cherokee wickiup is similar to the house my father was born on a reservation.
In "Indian Territory" - Now called the Oklahoma panhandle**

Still, in many ways, I may be one of the luckiest men I know. So was my father, but neither of us started that way. He was born in a Cherokee "wattle and daub" house called a wickup, with no running water, heat, or inside plumbing, on a Cherokee reservation in "Indian Territory," now known as the panhandle of Oklahoma. He was the first of five boys, but my father moved south with his parents to Texas's high plains before his four brothers were born.

This house is similar to my father's house in Cotton Center, Texas

His father set up shop as the town's blacksmith in Cotton Center, the population of something less than 100 in those days. My father was nearly 6'2" and always big for his age. When he was thirteen, his fourth brother was born, and his father, my grandfather, came to him with bad news.

"Son, we only have room and food for four boys. You need to move out and find a job." He worked as a ditch and fence post hole digger for the next two years, sleeping wherever he could with his co-workers.

His luck changed when he answered an ad for a "house boy" in the home of Meade F. Griffin, the District Attorney in Plainview, Texas. Meade Griffin saw the potential in my father and, besides teaching him manners and how to act around educated people, made sure he finished high school, then two years at the local junior college, and sent him on to the University of Texas where he earned a Bachelors in Business Administration (BBA) in 1936.

Meade Griffin did the same thing for at least two, possibly

three of my father's four brothers. The Thomas family owes him a great deal. Justifiably, he is a legend in Plainview for his many good works, but he was much more than that. He had an essential role as a called-up Army reserve colonel in preparing and executing the Nuremberg War Trials. Then he went on to serve 18 years on the Supreme Court of Texas.

When my father graduated from UT in 1936, Corpus Christi was amid a period of very rapid expansion. Its several new oil refineries and significant port facilities rapidly turned Corpus Christi into an important small city. My father took a job as an accountant at one of the new refineries. Shortly after he arrived, he met and married my mother, the "catch of the city" that year by all accounts and various old photos. She was 18 and beautiful, living in one of the most elegant homes in the city. It was a huge step up from Cherokee wattle and daub. I was born seven years later at 205 N. Carrizo, one of the better homes in Corpus Christi at the time.

My Mother's Parents' House at the time I was seven. The panoramic view of Corpus Christi Bay from the 2nd story lured me to the sea!

However, the fortunes of my Mother's family were failing. I was the second child, the first of three children born eleven months apart (AKA delayed action triplets). Our family of six had outgrown my Mother's parents' graceful home. We moved into a small house at 241 Indiana.

The move also made way in the elegant house for my Uncle George, my Mother's younger brother, and his new wife. My grandparents continued to live at the Carrizo Street home until the death of my grandfather in 1948. My grandmother moved to 521 S. Broadway on a bluff overlooking Corpus Christi Bay a few years later. It was a slightly smaller but even more elegant home near downtown. A long-time mayor of Corpus Christi had built it as his showpiece, and it was one of the city's best homes when I was born. However, by the mid-1950's it needed significant repairs. She rented out bedrooms to make ends meet. As an early teenager, I remember wondering why my family had two lovely homes, but both needed significant upkeep. Still, my father was living much better than he did when he was growing up.

The tension between my father, thinking he "had arrived," and my mother, who saw her standard of living slipping, eventually broke them up. After the divorce, my mother took a job as the administrative assistant to our local US congressman. She also ran for office as the president of the school board, which she won handily. However, less than two years later, the congressman was defeated, and she went looking for another job. She had attended a national meeting of all school board presidents in San Francisco and fell in love with it, so she focused her job search in the Bay Area. She soon found a job as head of public relations for "the world's largest" orchid farmer.

We moved to San Francisco with five days' notice. The move to SF was a lucky stroke for me. I was less than a month away from my 15th birthday and athletic. We played football year-round in Corpus, so I tried out for the local high school football team and easily made it. Indeed, I think I was the only member of the team that played every play. I played both offense and defense, as well as kickoffs and punt returns. At the end of football season, I tried out and made the wrestling team in the highly competitive middleweights, winning the league championship in the 175 lbs weight class my second year. Thus, I was very welcome in my new school. I am sure that I would not have been first string in either of those sports had I stayed in Texas. Football is a religion there, and high school football is highly competitive.

Also, the Bay Area school systems were at least a year more academically advanced than Texas, which left me much better prepared for life than had I stayed in Texas. I still consider the move to SF as a significant positive turning point in my life. I was lucky to have the opportunity to expand my horizons beyond Texas, but tragedy struck the day before New Years' of my junior year when my Mother was killed in a car wreck in Arizona as we returned from Christmas in Corpus Christi.

My younger brother, sister, and I were in the car, too. All of us suffered significant injuries, but we could walk independently and took the bus back to Texas to rejoin living with my Father. It was good to see my father again, but the circumstances did very much dampen the event. I quickly realized that I missed California very much and wanted to return as soon as I finished high school. It was nothing against my father or Texas; I just realized I liked the Bay Area and its lifestyle much better than that of Corpus Christi, Texas.

Two years later, at the end of my freshman year at Del Mar College, the local junior college in Corpus, I packed my bags in my 1953 Studebaker coupe, which I still think was one of the most beautiful cars ever made, and headed west.

As an experienced camper, I never gave a thought to hotels, motels, and the like. I camped all the way. Indeed, over the next four years, I made the 2100-mile trip from the Bay Area to Corpus eight or nine times, including once with Linda, my fiancé, and then once again eighteen months later with her on our honeymoon. That time it was winter, and we stayed in motels. I still consider it lucky that my father had given me the camping skills to make those trips. Linda had camped with her family and was a good camper, too. The trips, in and of themselves, taught me a great deal.

Once in California, I worked as a gas station attendant and enrolled in the College of San Mateo (CSM) for my sophomore year. I had spent my freshman year at Del Mar College in Corpus Christi. At Del Mar, we had a Daily Bulletin, an informal newspaper. I had helped publish it, so I knew how it operated. CSM had no way to get the news out to the student body except announcements over the cafeteria's loudspeaker, so I approached the Student Council and asked for the funds to set up a daily bulletin. After a fair bit of discussion, I got my money. The next semester I ran for Student Council, which I won. The semester afterward, I ran for Student Body President, which I lost. The semester after that, I transferred to the University of San Francisco (USF). The President of CSM got me admitted to Stanford, but I took one look at the astronomical tuition cost at Stanford and said, "Thanks, but no thanks!" I knew nothing about student loans in those days. And my best friend in Corpus had gotten into Stanford, but not Rice, so I did not understand Stanford's prestige.

My Mother's brother, my Uncle George, was the executor of my grandmother's estate, which was more substantial than I realized at the time. Indeed, I knew I could not access it until I was 30, so I asked no questions as it seemed inappropriate for me to ask. A year later, when I mentioned to him that I could have gone to Stanford rather than USF, his reply startled me. He understood how prestigious Stanford was, which I did not at that time, and informed me he could have easily covered my tuition from the estate and that he wished I had talked to him first. Oh, well! USF was a great school with one of the best records of any school in California in getting folks into law school, which I thought I wanted to do eventually. And San Francisco was a fantastic learning experience in and of itself for a semi-poor boy from Texas.

My brother Bob, "the brains of the family," joined the Navy in 1961, a couple of months after graduating from the University of the South with an honors degree in chemistry. He had applied for several jobs. But when potential employers learned he had not yet served his military obligation and was eligible for the Draft, they would not hire him, even though he had graduated near the top of his class. He decided to go ahead and join the Navy. Sometime later, he told me that he joined because I always talked about what fun a Navy career would be.

Bob became a Supply Corps officer. His first assignment was to a Long Beach-based destroyer escort assigned to radar picket duty in the North Pacific Ocean. At that time, I was studying history as a pre-law student at the University of San Francisco, and we linked up three times in two years. I visited him twice, once in Long Beach and again in beautiful San Diego, and he saw me once when his ship made a port call in San Francisco. I was particularly impressed with, and genuinely liked, his fellow officers, their life, and spirit. I was confident I had made the right career choice in wanting to be a naval officer.

During one of those visits, he pointed out that because the Draft made everyone serve two years, the pay was low, but all service members' salaries quickly doubled, e.g., $571 versus $298 for an Ensign when they went over two years in service. The doubling of the wage at the two-year point induced draftees completing their two-year commitment to stay in the military. The time in service could be on active duty or in the reserves. Bob suggested I join the Naval Reserve as soon as possible to gain the pay benefit of being "over 2" when I did come on active duty.

So, I went down a few weeks later, on 20 July 1965, between my junior and senior year of college, and enlisted in the Naval Reserve. It seemed like an excellent idea at the time (and indeed, it turned out to be in the long run), but it had significant repercussions for me within the next 19 months. The recruiter had assured me that I would not be activated until I had completed my degree if I were a full-time student in good standing.

However, 19 months later, as Vietnam heated up, I was yanked out of college and called to active duty as a seaman. I had completed all the required courses to graduate but still needed eight more credits (less than three complete courses) to meet my degree requirements. In the end, spending nearly two years as an enlisted man on sea duty was a real blessing, but it certainly did not seem like it at the time.

I had taken the first half of boot camp in the summer of 1966, between my junior and senior years. In February 1967, with one part-time semester to go, I was ordered to report to Treasure Island Naval Base San Francisco within ten days. From there, I was bused to recruit training in San Diego to finish boot camp. I was pretty low, but I had faith that it would all work out.

I did not like boot camp very much, but one incident still makes me smile. That incident involved rifle training. Being from Texas, I had wanted to shoot for as long as I could remember. My father had taken me to the police gun range every Tuesday night since the month I turned 11 years old, the minimum age to be allowed to shoot there. I had gone through their training course and qualified for the marksmanship merit badge as a Boy Scout. Archery was another passion of mine at the time, and I had also become certified as an "American Archer" at the minimum age. When I was 16, I worked as a counselor teaching archery at a scout camp in the beautiful hills behind Monterey Bay, south of San Francisco.

Our rifle instructor was a US Marine sergeant-major sniper instructor on permissive temporary duty. I was in awe of him. He and I had spent a lot of time after-hours trading instruction on archery, which the Marine got a big kick out of, and me shooting on the rifle range under his expert instruction. We had unlimited ammunition thanks to the USMC and my shooting, already good, went up dramatically. So, when I went to the boot camp rifle range with my fellow "boots," I felt right at home.

"Load one round," "Aim," "Shoot" the orders came down five times. Then "Load five rounds" "Shoot at the target as fast as you can." Finally, they put up a new target and said, "Load five rounds. This time you are to aim and try to hit the center of the target each time you fire. Shoot at your own pace." I picked up the rifle and, in less than 20 seconds, fired five bullets which hit the target in a tight little group that touched the center of the target. (The sights were slightly off, but the five rounds made only one ragged small hole.) I set the rifle down and waited for further orders as we had been instructed to do. Less than a minute later, the instructor came by and saw my rifle lying down. He assumed I had not yet shot, so he told me to commence firing.

"I am out of ammo, sir."

"You shot all five rounds of aimed fire?"

"Yes, Chief."

"Let me see your target!" and he reeled it in.

Looking at the tight little group of only one hole near the center of the target, he said something about it being maybe the best shooting he had ever seen from a "boot" and asked: "How would you like to be a SEAL, son?"

"Chief, I want to fly." "Good luck, son, but keep the SEALs in mind. They need folks who can shoot."

At that moment, I had no idea who the SEALs were, but I indeed did find out later in my career. It had been fun shooting over the past nearly ten years, and I did volunteer to be part of USS HORNE's "Landing Force." There I got to shoot the Browning Automatic Rifle (BAR), a small machine gun, and the M-14, the follow-on the M-1 of World War II fame, as part of our training.

However, since that time, I have not shot a rifle hardly ever again except to qualify as an expert while I was with the Air Force a dozen years later. I never got the chance to shoot a rifle on a range with the Navy, but I did shoot on the Naval Security Group pistol team for two years. I used to love to shoot a rifle, but I never had the opportunity to shoot when I became an officer. Just too busy doing other things! Odd! (& sad!)

After six or so weeks, our instructor told us to fill out our "dream sheet." What did we want to do? Ship? Shore? And where? The US? Overseas? You could also write a few sentences as to why you wanted what you wanted. I filled up the page. I asked to be assigned to one of the bases in or near San Francisco so I could finish my degree and put in for officer candidate school. I mentioned I was an Eagle Scout with the God and Country plus the Explorer Scout Silver awards also, that those awards made me one of the most decorated Scouts in the State of California.

A 2nd class Yeoman interviewed me, and I explained I had not flunked out of college but was pulled out with less than a semester to go. I mentioned I was also working 20 to 30 hours a week, was singing in several organizations (which was a significant time commitment), and editing the daily news bulletin (also a large time sink, with deadlines). Moreover, because of my time as a Boy Scout, I thought I understood leadership's demands and could handle the load. I summed it up by saying my time as an enlisted man would make me a good officer. He bought it and said he would try to help me. The next day they posted our assignments, and I had orders for a cruiser. I was crushed as flat as a pancake. However, it was, in reality, a blessing in disguise, but I certainly did not know that for some time.

I saw the 2nd class the next day at chow and confronted him. "You said you would help me, and I got orders for a cruiser! What happened?"

"Hey, jerk! That cruiser is being built in San Francisco and will not leave there for at least nine months. You should have more than enough time to complete your degree! And you will be getting a dollar a day sea- pay which will nearly double your pay!"

    Suddenly the sun came out, and the gloom disappeared. A choir of angels broke into the Hallelujah chorus as a great weight lifted from my shoulders! I was going back to SF, and that $1 a day did nearly double my pay of $35 a month! When I went over two years of longevity that summer, my income went up to $71 a month! $100 a month was about what I was making in college, working half-time at a dollar an hour. And in the Navy, I had free room and board! Life was good! Someone was looking out for me, no doubt about it.

USS HORNE
A Brand New Missile Cruiser
Then
Training, Training, and More Training

Chapter Three

One of the things I sensed immediately upon reporting to the USS HORNE's pre-commissioning crew was that I would be learning a great deal to help me in my chosen career. This became more and more evident as I spent more time on HORNE. We completed her construction, did dockside checkout, took her to sea, showed her off in several foreign ports, and then to combat off Vietnam.

Indeed, I now realize I probably would not have been selected to be an officer had I not served on HORNE and caught the eye of the commanding officer, newly promoted Captain Stansfield Turner. He had "ADMIRAL" written all over him even then. He had been Commandant of Midshipmen and a football star at the Naval Academy, a Rhodes Scholar, and selected early for his last three promotions. He was obviously a rising star in the Navy. He would retire as a full admiral with four stars after heading the CIA for his Annapolis classmate, Jimmy Carter.

Indeed, years later, in early November 1980, when I was a lieutenant commander, he told me that should Jimmy Carter win reelection, he would be "returning to the Pentagon in a more senior position," which had to be Chairman, Joint Chiefs, and he would have a place for me. Several years later, his son Geoff said much the same thing. "Carter had promised my dad the position of CJCS." The right person with which to have made a good impression! More on this later.

On HORNE, I was initially assigned to the deck force. As a Sea Scout, I knew the drill, but it was not what I had wanted to do. I had asked to strike for "quartermaster," the men who ran the navigation operations of the ship itself under the watchful eye of the ship's navigator, and the Officer of the Deck, the man controlling the ship's movement when underway. Quartermasters had to have a great deal of seamanship and navigation knowledge, things I had enjoyed studying as a Sea Scout. However, the ship's administrative office was looking for additional people. In reviewing my record, they saw that I could type. Congratulations, Thomas! You are now a yeoman (Navy for clerk). It beat chipping paint as a "deck ape." Plus, the chief in charge of the admin office was sympathetic to my situation when he heard my story. He allowed me to leave the ship early to attend the three classes I needed to finish my USF degree. Even better, they were all music classes that had no homework. Music is another passion of mine. I have even sung some professional opera after I retired from the Navy. Once I had my degree, I applied for Officer Candidate School (OCS).

USS HORNE DLG-30

As the Vietnam War escalated, the nation needed to significantly increase its armed forces. The competition for slots at OCS, which certainly beats being drafted to be an infantryman and sent to the rice paddies and jungles of Southeast Asia, dramatically increased. So, I was not surprised when my application was rejected. I did not have a technical degree, and I was not at the very top of my class. In looking back, this refusal was another good thing that did not seem so at the time. I still had much to learn on USS HORNE and knew it.

I fondly recall many things from my time at HORNE. The absolute high points were my conversations with Captain (later Admiral) Turner. I became his driver, at least in part because I knew San Francisco, and in part, because I was an Eagle Scout and interested in a career in the Navy as an officer. The fact that I was an Eagle Scout with two other high scouting awards gave me a fair bit of initial credibility with Captain Turner as well. He was intimidatingly smart with a genuinely fantastic memory. I thought my Jesuit professors at USF were bright, but his mind was at a completely different level. He was so intelligent and sure of himself that he treated everyone with the courtesy and respect they deserved. He wanted to know what everyone he met thought and knew and treated everyone as an equal. (Which is, I now know, the mark of a great man.) We had many fascinating conversations as I drove him to and from his home to the ship (about 30 minutes each way) and to various meetings during the day. It was what talking with Socrates or Aristotle must have been like for their students. I learned a great deal from Captain Turner.

I most clearly remember his conversation, which dealt with the Israeli Destroyer Eilat's sinking by a Russian Styx surface-to-surface missile. An Egyptian patrol boat fired the Styx, but it had never left port. He told me that he did not understand how the Egyptian vessel could acquire its target from so far away. I asked him if it was possible to use the radar in an airplane and send the target information via a data link. Not long afterward, he let me know that Naval Intelligence had suggested the same thing as my idea of the airborne data link. He genuinely liked that the same thought had come from a seaman, which I was at the time. He was so gracious and intelligent that I was not surprised that he made four stars, full Admiral, in near-record time.

About two months before Captain Turner left, he suggested I resubmit my application for OCS before he left because, as he so wisely put it, "You are going to want my signature on your application. My relief will not know you from Adam." (And he was going to an important billet, that of Military Assistant to the Secretary of the Navy). At my retirement ceremony 20 years later, he told the story of the extra steps he took to get me selected. I also know that for the next 20 years, I tried to pay him back by being as good an officer as I could be.

The last time I saw him was in 2000. He was the visiting scholar at the Naval War College (NWC), where he was its President 24 years earlier. He is famous for turning the NWC from being a "fluff" assignment to the US military's premier institute of advanced education and thought. Indeed, the other services copied his academic initiatives in revamping their schools of higher learning. After I retired from the Navy in 1988, I returned to Newport as a civilian contractor to install the NWC's first local area network (LAN). At that time, Admiral Turner had been gone from the NWC for 14 years. However, many of the fine professors he recruited were still there. They privately referred to themselves with great pride as "part of the Turner Revolution."

When I last saw Admiral Turner in early 2000, I was recently assigned, for the 4th time in my career, to the NWC. This time I was Johns Hopkins University's Applied Physics Lab (JHU/APL), liaison to the NWC. APL was the Technical Direction Agent of the Navy. My assignment was to be the two-way conduit between the Navy's technical and strategic think tanks, a plum assignment. At that time, I thought it would probably be the best job I would ever hold. He pronounced himself exceptionally proud of me and said he never doubted I would be an excellent officer. To me, it was the highest praise I ever received.

In January 1968, HORNE moved its homeport from San Francisco to San Diego and, after about five months of training and systems tweaking, deployed to the Western Pacific. I was off on an adventure, still waiting to see if I was accepted for OCS. Indeed, my 2nd OCS application left HORNE during our ten-day stopover in Hawaii. We were there in Hawaii when Robert Kennedy was assassinated in Los Angeles. I will never forget the grief that seemed to overwhelm the islands at that time.

From Hawaii, we went to Japan to show off our new ship, the most advanced in the Navy at the time. Among many other new capabilities, it had a powerful, for its day, computer system built into it from the start.

Our Executive Officer (XO), the second in command, was Commander Kleber Masterson, later Rear Admiral (RADM). He had just completed one of the first master's degrees in computer science awarded by the Naval Postgraduate School. He was most interested in trying new things with the central computer built into the ship. He also started several computer science classes, and I enrolled in his Computer Science 101 and Fortran courses. Having studied computer science too, turned out to be an excellent thing later in my career.

As we left San Diego en route to the Gulf of Tonkin, my duties changed utterly. Because I had a completed background investigation (BI) in my file as part of my OCS application and thus could be given a security clearance immediately, the Captain decided to assign me to the intelligence team as their assistant. I was given a Secret clearance.

My workspace became the underwater plot (UW) corner of the combat information center (CIC). UW Plot doubled as the Intelligence Center when there was no underwater threat (submarines or mines). This assignment was a large part of my early education in the Navy. What I learned in the next six months was an immense benefit for the rest of my career. Another lucky break!

Indeed, being in CIC was a huge plus. Naturally curious about technology, I made a point of learning the function and how to operate every piece of equipment there during the routine operations as we transited the Pacific Ocean en route to the western Pacific. There were multiple radar and radio systems in CIC; however, the WLR-1, the electronic warfare (EW) receiver, the ULQ-6, and the ALQ-91 jammers captured my attention. I was just drawn to them for some reason, probably because I understood their functioning from a course I had taken in college on music's physics. I am not the first person to realize that the physics of EW and music are the same in many ways.

The ship was also one of the first to have been built to accommodate the Navy Tactical Data System, which fascinated me. I pestered the data systems technicians with many questions. I was probably the most interested person in the crew. I certainly asked the most questions. They were enormously proud of their new system, the best in the Navy at the time, and were happy to talk about it, so we became good friends. The information I got from those discussions on that early data system has also stood me in good stead for the rest of my life.

From Japan, we went to Subic Bay in preparation for deploying to the Gulf of Tonkin. Before we left Subic Bay en route to the coast of Vietnam, we embarked a Naval Security Group (NavSecGru) direct support (DirSup) detachment, a team of "spooks," on board. They worked in "Supplemental Radio" (SupRad) close to CIC.

They were extremely cautious in describing their duties, especially the enlisted men who were obviously highly intelligent. However, I had been in their spaces before they came on board. It was full of electronic equipment, so it was easy to guess what they did. During my five months off Vietnam, I worked very closely with them because they also needed the intelligence I pulled off the classified fleet broadcast. I made a clipboard of the highest priority messages that I took to the Captain. He told me who to be sure saw messages of particular pertinence or interest. Once the specified officers had seen the messages, the clipboard was available in my "office" there in CIC. I was the day watch Intelligence Officer, but I saw myself as basically a glorified clerk.

However, there was a huge upside. My job was to read all the intelligence and operational message traffic that came into the ship. It was fascinating. I was also learning a great deal about the operations, capabilities, and problems of many systems. I was often able to answer questions that came up in CIC from my readings, which made me a valuable and welcome part of the team. E.g., our air search radar detected an aircraft flying at over three times the speed of sound (Mach 3.2) at over 80,000 feet coming up the center of the Gulf of Tonkin. It disappeared over south China heading north. "What the hell was that!" was the question flying around CIC. I must have been the only one who had read anything about a brand new, Top Secret reconnaissance (spy) plane, the SR-71, and was able to bring everyone up to speed quickly. It caused quite a stir, and it did not hurt my credibility.

Our role in Vietnam was the first line of air defense for the aircraft carrier battle groups operating further offshore and the cruisers and destroyers operating on the gun-line inshore. Our primary weapons system, by doctrine, were the F-8 and F-4 fighters we controlled. We also had the Terrier Surface-to-Air missile system, with a maximum range of about 40 miles, the 5" 54 rapid-fire gun, with about 12 miles of range, and two 3" 50 guns with an effective range of something like 6 miles.

The last was "my" gun. My time on the rifle range paid off here, too. I shot one of the highest qualification scores ever (68 rounds on target out of 80 fired) with our 3" 50 on the San Clemente Island Ship Gun Range. A passing score was 20/80, and many ships struggled to get that. My marksmanship with that gun also really boosted my standing with Captain Turner and the officers' wardroom. (It was also a lot of fun!)

I was strapped to the sizable optical gunsight with handlebars via a harness. It was like being lashed to a big power lawnmower, except instead of cutting grass, I was shooting 3.5" artillery shells when I squeezed the "throttle." As the ship twisted and turned, I used the handlebars to move the whole sighting mechanism up and down and side to side to keep the sight locked on the target, be it a target sleeve being towed by a plane, a target barge being towed by a tug or a pile of cars onshore. The gun followed my every move and stayed locked on the target if I did my job right.

When the target was in the center of the sight, as marked by a small circle, and I had permission to fire, I squeezed the trigger. BOOM! The magazine of the cannon held four rounds. If I held the trigger down, it went BOOM! BOOM! BOOM! BOOM! But I generally let up long enough after two shots to check to see if the target was still right in the middle of my sight before I squeezed the trigger again. This was necessary because the concussion from the gun's muzzle was knocking me around, and there could be smoke obscuring the target for a second or two. During that short pause, which also let the gun crew load two more rounds, so I could keep firing. It was all great fun, and I had quite a reputation among the ship's officers. However, even with ear protection, my ears rang for some time afterward.

I was in CIC for many airstrike support operations against North Vietnam. While all the operations were fascinating, there were several that made significant impressions on me. The most exciting event happened by seeming accident, but it led to the development of a totally new Navy capability. HORNE was assigned to the North Missile Air Defense position off North Vietnam. Our radars detected a pair of MiGs transiting from north to south just inside the shore. They appeared to be heading into the absolute maximum range of our Terrier missiles.

As we tracked them coming down the coast and into the reach of our missiles, we prepared to launch a pair of them at their maximum range. The guided-missile cruiser USS LONG

BEACH had just come into the gulf with its primary radars inoperable. It was now 15 or so miles to our southeast as it worked to fix its radar and bring it online. The fact that its unique radar was not radiating meant that the North Vietnamese did not know the LONG BEACH, which had the Talos missile system with a 70+ mile effective range, was anywhere in the area. We were broadcasting the MiG's position via the Naval Tactical Data System (NTDS) to all Navy units in the Gulf of Tonkin, including the Long Beach, indeed, especially the Long Beach.

The LONG BEACH brought up its fire control radars using our data to attempt to lock on the MiGs with their fire control radars with their very narrow beams, which are much less unique and also harder to detect. It was a challenging problem for LONG BEACH. Still, as we received permission to fire from the battle group commander, the LONG BEACH reported that she too had succeeded in achieving a fire control radar lock-on. She also had a firing solution. The MiGs were about 40 miles from us (our maximum range.) and 55 from LONG BEACH, well within the 70+ mile range of its Talos missiles. "HORNE, Hold fire! "LONG BEACH, Birds free."

As these words came over the radio, I watched the radar scope and saw two missile tracks appear from our southeast, where the LONG BEACH was located, and streak toward the MiGs. LONG BEACH had fired before they were given permission! "Roger, Understand Birds Free!" came the belated call from LONG BEACH. All is forgiven if you are successful and the TALOS missile tracks merged with the MiG tracks, and both disappeared. Success!

Years later, when working at JHU/APL, I had an elderly gentleman, an APL technician working on the Talos system on LONG BEACH at that time, tell me the story from the LONG BEACH view.

He also described how that event gave birth to the Collaborative Engagement Capability (CEC), a prime weapons capability developed by APL. CEC is now on all our guided-missile ships. He did not fully believe me when I told him I had been in that engagement until I detailed my very minor role in that historic event. I have also read that this event was not an accident, but rather, was planned. That may have been true, and I was not cleared for that event.

However, I religiously read all the SECRET message traffic every day as part of my regular duties, and I never saw anything about the planning for such an event in message traffic. Also, everyone in CIC on HORNE indeed reacted as if it was an accidental event. However, I must admit that there are levels above Top Secret that I was not cleared for at the time, so it is possible that the action was pre-planned at a high clearance level, and the CIC crew on HORNE had some excellent actors in it.

The NavSecGru detachment also had a comms link with the shore that was classified higher than Top Secret. It was possible that a link was used to coordinate the ruse, but it was a surprise to the great majority of us in CIC that day.

After each month-long line period off Vietnam, we visited a different Far East port: Manila, Singapore, and Taipei, Taiwan, then Hong Kong where, in August 1968, I learned that I would be leaving HORNE to go to Newport, Rhode Island, for Officer Candidate School (OCDS) in October. It was a dream come true, but it got better.

Not long after that, the officer in charge of the spooks, LT John Thomas (no relation), suggested that I apply to his organization, the Naval Security Group (NavSecGru). He noted I seemed to be well suited for intelligence work. LT Thomas pointed out I had an excellent memory for detail, was conscientious, and seemed to enjoy the work I was now doing. He was sure I would like the work of the NavSecGru Direct Support (DirSup) teams, even though he could not describe in detail what they did. I had a good idea of what that was.

I was sure he was right. I would like that type of work. He

also mentioned that the NavSecGru had a program that put people on submarines on special missions with very high visibility assignments, and I just might qualify for them. They also had officers who flew in reconnaissance aircraft on missions essential to national security. The selection processes for both programs, submarine, or air, were arduous, and the training pipeline for either program was long. Still, it was the best job in the US military, in his opinion. I was sold! (& I have come to completely agree with him!)

He gave me the name and address of the NavSecGru detailer (assignments officer), suggesting I write him a letter describing what I was now doing and volunteer for duty in NavSecGru DirSup. I did so immediately, and that too was a turning point in my life. The answer came back from the detailer that they were interested and would interview me at OCS during the "career day."

That was when recruiters from all Navy organizations came to OCS to pitch their part of the Navy to the fledgling officers. Most of the recruiters already had a list of who they wanted to recruit. The nuclear Navy guys had first pick. They were looking for engineers near the top of their class at the most demanding engineering schools. They had their list of names they wanted to recruit in priority order, just like a pro sports team draft. I looked forward to that day as the pitches were sure to be remarkably interesting. I had wanted to be a pilot, then a surface officer, and now I wanted to be a DirSup officer in the NavSecGru.

In mid-October 1968, USS HORNE's bell rang twice: "Bong-Bong," followed by the announcement "YN3 Thomas, Plankowner, departing" over the ship's public address system as I was "bonged-off" HORNE (the Officer of the Deck had the ship's bell rung in my honor), and I walked down the gangplank there in Subic Bay, Philippines Islands. I was leaving my first ship for the last time. I felt incredibly proud, somewhat sad, and undeniably excited.

My adventure was starting in full, but the previous 20 months had been very educational and exciting, and I was a changed man. Just blind luck had given me the chances I had just been given. I experienced the same feeling once as a defensive lineman in a high school football game when I picked up a bouncing football just dropped by an opposing back, and there was nothing between me and the goal line. I did not fumble then nor during my Navy career, but it seemed it was a near run thing several times for various reasons in the next 20 years.

I took a no-frills USAF bus for the nearly two-hour ride from Subic Bay to Clark AFB and boarded a chartered commercial airliner full of servicemen returning to the states from duty in Vietnam. It was one big party. After a refueling stop in Yokota AFB, Japan, we arrived at Travis AFB, about 30 miles north of San Francisco, where my wife lived. I took a Greyhound bus into the terminal in downtown SF. My wife had gone to school at Berkley while I was at USF. We had rendezvoused at that terminal many times, so this was familiar territory.

From there, I took the local bus to our apartment. That ride, on a familiar route, was a shock. As required by regulations, I was still wearing my Navy "cracker-jack" uniform, not unlike the Sea Scout uniform I had worn ten years before, but this time the reception was vastly different. You would have thought I had leprosy. People shunned me everywhere. And many were overtly hostile. Even a little old lady came on the bus, saw me, and then moved to sit down next to a drunken bum rather than sit with me. So much for "gratitude for defending your country."

My wife was also in shock. She was sure, I now know, that I would not be selected for OCS and would come home and go to law school, my "fallback" plan. Now that I had been selected for OCS, she bravely got ready to move to Newport and assume a Navy wife's role. However, it was clear that the attitude of the people of northern California's attitude of overt hostility to the military was harder on her than it was on me. In my hometown, Corpus Christi, Texas, a military career was at least the equal of any professional career such as a lawyer or doctor. This was clearly not true in the SF Bay area, her home. I do not believe she ever fully understood that I considered a naval career to be the best possible one to which one could aspire. Still do. Indeed, I had also considered the merchant marine as it is my firm belief that any career at sea, be it military or civilian, beats any career on land. However, I now know the most challenging job in the military anywhere is that of the wife left behind. That was certainly true of the Navy wives as I came to know very well, but as a young man, embarking on a dream of a career, you do not think of such things. You think of ADVENTURE!

I left San Francisco three weeks later for the first two months of OCS, where I would have no free time. It was late October, and I came back to SF over Christmas to pick up my wife and bring her to Newport. On 2 January 1969, we took the train from Oakland to Chicago and then went to Providence via air. The -16F temperature shocked us when we got to Chicago. Still, we bundled up in a warm taxi and made the transfer from the train station to the airport as fast as possible.

Newport was another trial, especially for my wife, who had rarely seen snow in her life. The stiff ocean breeze magnified the cold, and it was the coldest winter since 1938. Snow drifted up to the 2nd story of our new barracks, and Narragansett Bay froze out as far as we could see. A full-sized dump truck with a snow blade doubling as a snowplow got stuck in front of the officer cadet barracks. It was several weeks before it was freed. The first day above freezing for 24 hours was graduation day, 11 April 1969. It was a hard time for everyone, but we were now officers and free to go.

My interview with the NavSecGru detailer on "career day" at OCS had gone OK. They were looking for someone with administrative experience, and as an ex-yeoman, I fit their needs. I was assigned to the Naval Security Station, Washington, DC, the support staff for the Naval Security Group (NavSecGru) Headquarters, as an assistant admin/personnel officer. I had hoped to be directly assigned to the NavSecGru Direct Support (DirSup) school pipeline, however the needs of the Navy were that the NavSecGru needed an admin officer at its headquarters support command, and I had the essential qualifications they wanted.

They assured me I was in a holding pattern, waiting for the schools I needed to qualify for submarine DirSup. The trade was that I was now working in the organization that I wanted to be part of and given the coveted "SCI" clearance. On arrival, I made my wishes known to my new boss, Lieutenant Commander (LCDR) Orville Heinz, an up-through- the-ranks admin type. He understood my desire to be a submarine DirSup officer and made a deal with me. "You are assigned here for three years. Give me your best for a year. I will get you assigned to the communications department for career-broadening for the second year. We will then try to get you assigned to the DirSup training pipeline a year early."

I did one thing of note during that year that I am still proud of to this day. As a junior officer, I served twice as a "Casualty Assistance Officer," the officer who accompanies a clergyman to tell a family their son or daughter was not coming home. My job was to set up the funeral arrangements per their wishes. In both of my cases, the parents of the deceased were divorced. The immediate family members assured me the deceased would be revolving in their graves if they knew the father, in either case, was getting a dime of the life insurance in question.

The assignment of the beneficiary of that insurance was automatic at that time. If the service member was unmarried, the parents split the insurance. I submitted a change to Page 2 of the service record, where the service member could specify life insurance beneficiaries. The new change made it clear that both parents would split the insurance if the service member was unmarried, and no selection was made. My change also allowed a service member to name one parent or the other or receive their life insurance, the serviceman's choice. The change to the Page 2 was accepted. I felt good about that, but I could do nothing for my "clients" ex-post-facto.

After that year, I was assigned to communications watch officer (CWO) duties as promised. I learned several things there, among them how to set up several different encoding/decoding machines. I also set the crypto for the Chief of Naval Operations' (CNO) official residence at the Naval Observatory two days out of every eight. CWOs were also the offline comms guard for the CNO and the NavSecGru HQ. The highest classified messages were offline encrypted and had to be decrypted via hand encryption/ decryption machines. There were several different types, and I understand these were among the systems Johnny Walker, the infamous Navy spy, gave to the Russians. He should have been roasted alive!

More on Walker later!

There was one type of hand-encrypted message marked "For OFFICER EYES ONLY" (OEO). It came in on one of the most challenging systems to set up. All the CWOs generally had to ask the watch chief to help us set up the crypto machine and then leave so we could finish the decryption in private. It was all demeaning to the chiefs, and thus embarrassing to me.

When Admiral Zumwalt, the new, young head of the Navy, announced he was looking for ways to give more responsibility and recognition to the chiefs, I had an idea. Why not change the OEO marking to be inclusive of the chiefs? I talked it over with my new boss, LCDR Harry Ohan, another up-through-the-ranks professional. He liked my idea of changing the OEO marking to "Special Security Officer Special Category (SSO SpeCat). I was pulled off the watch, made his assistant, and given the job of running this idea through the Navy bureaucracy. I still do not like bureaucrats to this day, but we succeeded, and SSO SpeCat became an official classification, and senior enlisted men could now see these messages.

True to their word, my bosses released me in March 1971,

fourteen months early. I was ordered to Monterey, California for Russian Language school, the start of the submarine DirSup pipeline. During my 22 months in DC, I also took three courses in computer science. Additionally, I sang with Norman Scribner at St Alban's Episcopal Church on the National Cathedral grounds for almost two years.

Norm was only a few years older than me, and we became good friends. I was delighted to learn years later that he was considered a national treasure of the arts upon his retirement in 2013. We did have one other interface, in October 1971, that I will describe shortly. The computer courses also proved to be exceptionally useful in my future career.

At language school, it was tough going. I could not even see the difference between some of the Cyrillic letters. After several months of struggle and frustration, one of my instructors suggested I get my eyes examined, and, sure enough, that was the problem. I put on the new glasses, and it was like a dirty window had been opened. I was going to be rolled back a class, but the Navy needed folks in Japan ASAP. Because of my operational experience as an intelligence type as an enlisted man, they decided to ship me out with the class in front of me instead of rolling me back six weeks.

I took the language qualification test to see how much more I needed to learn, which I knew to be a great deal. To the surprise of all, I qualified as a linguist, and off I went to Submarine School, then the WLR-6 course, a series of classes on how to operate every reconnaissance and sensor system on a fast attack submarine, and finally, another series of classes at Fort Meade, Maryland. Those courses include cryptanalysis, traffic analysis, special security officer responsibilities, communications school, and DirSup Officer responsibilities (dos and don'ts).

The thing I remember most about Submarine school, other than all the interesting equipment we got to play with, was the escape tower. It was 50 feet high, and we entered it via an airlock at the bottom wearing submarine escape gear. The gear consisted of a flotation device and a breather. Our job was to just go through the lock, inflate our flotation device, and "Blow and Go." You needed to be breathing out the whole way up to keep the airway open to allow the compressed air in your lungs to escape. Once you started "blowing out," you were to release your grip on the rails at the bottom of the tower. You come up like a rocket, but you must keep blowing out the whole way, or you could burst your lungs as the air in your lungs expands as the pressure on them decreases, and they expand as the depth decreases.

I entered the bottom of the tower and looked up to see what it looked like from way down there. Hands reached out and grabbed, dragging me back into the airlock. "You panicked and froze!"

"No, I was just curious as to what the tank looked like from the bottom."

"Your delay could cause folks behind you waiting to enter the lock to die!"

Got it! My bad!"

"Are you ok to go again?"

"Sure!"

I climbed back into the bottom of the escape training tower, and up I went. It was fun! I put my arms and hands down by my sides and tilted my head back to make myself as streamlined as possible, and flew up the tower, popping up like a rocket. The instructors afterward told me I had come further out of the water than anyone they had ever seen. My body surfing off San Francisco had paid off! Many folks did not like that drill, but I thought it was fun and would do it again anytime!

Back to Norman Scribner: While I was at the series of courses at Ft Meade, I happened to notice a newspaper story saying that Leonard Bernstein, the famous conductor of the New

York Philharmonic, and composer (West Side Story), had selected my friend, Norman, to be his stand-in for the full-up performance test of the newly built Kennedy Center for the Performing Arts. Norm would conduct the very first performance in the recently completed Kennedy Center as a full-up test of the Center before Leonard Bernstein did the formal opening with the "Who's Who" of DC in attendance. Norm elected to perform Bach's B-Minor Mass, a massive, challenging work. Having sung that piece, I can appreciate the work that goes into its preparation for presentation.

This performance was the full-up test of everything at the Center, from the ticket sellers and takers to the ushers at the front of the house to the back-door security and everything in between. They were even going to test/train the kitchen and the wait staff with a formal reception in the roof garden afterward.

I immediately phoned Norm and congratulated him. He apologized that we could talk for only a moment because he had a million demands for his time since the great news had broken. Indeed, he was off to an interview on one of the major TV stations at that moment. He did say he had a few tickets to the event and the reception afterward, and he would leave a pair for me at Will Call. Maybe we will have a chance to talk there. I figured that was a very low probability but was delighted to get the free tickets to this historic concert, both costly and in high demand.

I asked my classmates there at NSA if any of them wanted to go with me. There were no takers, so I went by myself. I was not going to miss this concert! I arrived at the Kennedy Center's Will Call on the appointed date and time, collected my ticket, and the coveted invitation to the after-performance reception in the rooftop garden. I proceeded to my seat, which was terrible! It was literally behind a pillar, which became a scandal later, and those seats were subsequently removed, but I was stuck for the moment. I noticed that while most of the hall was packed, there were a few empty seats in the Dress Circle. I decided to snag one of them at the intermission, as I had often done in college at various San Francisco venues.

Near the end of the intermission, I went up a side stairway that I was sure would lead me to the Dress Circle. However, it only led to another longish hall which I hurried down. I was a little worried at that point that I would arrive after the concert had resumed, which would be rude, especially as an interloper. I opened the door at the end of the hall and walked into an empty 20 by 20-foot room. Now totally bewildered as to where I was, I hurried across the empty chamber and opened the door on the far side....to the President's Box, which was unoccupied. I understand the adjoining empty room I had just hurried across is now a "Green Room" for entertaining and relaxing. Everyone in the Dress Circle and box seats had seen the door to the President's Box open and had turned to see which famous member of the White House staff had decided to attend. Maybe it was Nixon himself!

The seats were in place, but there was no carpet. No problem! I sat down. Nice seat! A few minutes later, two fit guys in their early '30s, with earphones in their ears, sat down on either side of me. Obviously, the Secret Service! "Who are you?" one of them asked. I replied, "I am a naval special intelligence officer who used to sing for the conductor, and I just accidentally stumbled in here." They asked for some identification, and I produced my NSA badge and the invitation to the reception.

I may well have been the only one in the audience with NSA credentials and an invitation to the reception, were not easy to get unless you knew someone, as I did. So, in those calmer days, both Secret Servicemen relaxed and moved back a row. We all enjoyed the concert, and then I headed to the rooftop reception, where the fun began in earnest.

Because my seat in the Presidential Box offered a quick route to the rooftop, I arrived as the receiving line formed with Norm at the head. I was the first in line. He was charged up and gave me a big hug as I started down the line. "Thank you for coming and sharing the best day of my life!" was our only exchange that night, and I moved on down the line.

However, the next hour-plus was a complete hoot. Every little old lady and social climber in DC was there, and they all fell all over me, pretending to know me. I must be SOMEBODY! I sat in the President's Box BY MYSELF except for the security detail "guarding me," and I obviously knew the conductor very well. "Don't you have a house in the Hampton's near mine?" was a typical type of question.

I just played the role and admitted to nothing, as I had recently been trained to do at Ft Meade. (No joke. We could not admit who we were even at NSA and were given suggestions on deflecting questions.) Tonight, it was great fun and great practice in a benign environment. I still chuckle when I think of it. One of the most fun concerts and parties I ever attended!

Once we graduated from the DirSup Officer course, the six class members were ordered to their operational duty stations. Three went to Rota, Spain, and three to Misawa, Japan. The three of us going to Japan were offered the opportunity to select our means of going there, either by ship or plane.

Bill Brinkmann and I both chose to go by ship and sailed on the SS President Wilson, a member of the US merchant fleet about to be retired. It was a great trip on a historic vessel. It was also the last trip for the crew, and they went all out to make it memorable. We sailed from Oakland to Ensenada, Mexico, then Hawaii, where we had a full day to explore.

Now familiar with Oahu from my time there on HORNE, my wife and I rented a car and, with our 18-month-old son, went all the way around the island. From there, we went straight to Yokohama. We did have to divert south to skirt a storm, but it was a trip to remember for the rest of your life. In port, we were met by a USAF school bus for another "no-frills" ride across southern Tokyo for 20+ miles to Yokota AFB. We took a flight the next day to Misawa Air Force Base, our home for the next four and a half years.

Once in Misawa, our first job was finding a place to live. In alignment with my dream from 1958, we elected to buy a small house off base to experience Japanese culture first-hand. It was a wise choice, and when we left Japan six and a half years later, we took $35,000 with us from its sale. (A LCDR's annual salary, with flight pay at the time, was about $24K.)

The next thing was to start learning exactly what we were to do. Naval Security Group Activity Misawa had several missions. It was a mainstay of Navy signals intelligence. It also provided trained technical intelligence collection (DirSup) teams to man ships, aircraft, and submarines engaged in "special missions" against Russia. Its detachment in Pyong Teck, Korea, also supplied personnel for direct support for missions against North Korea.

Generally, the team members are trained for either air or submarine operations. From this pool of trained specialists, the surface detachments were also formed as needed. This was precisely where I wanted to be because I wanted to do all three. But, after my time at sub-school, I mainly wanted to participate in special mission submarine operations, which clearly have the most visibility in the Navy for those with ultra-high security clearances, those with three or more steps beyond just "Top SECRET." Indeed, many maintain special mission submarine operations are the brightest jewel in the entire US intelligence community's crown, bar none.

Interestingly, in the following years of my career in intelligence, I found that most Army and Air Force intelligence professionals had no idea of how significant submarine intelligence operations were, and I certainly could not tell them. It may still be the case, but I have not been cleared for these operations for many years. At least three books describe submarine intelligence operations in far more detail than I intend to go into here. All three stress how exceedingly important these missions were.

In reality, I had been focused on submarine special mission operations since I first heard about them while on USS HORNE nearly four years earlier. And now I was on the threshold of realizing that dream, or so I thought. I had not been in my new office for two hours when I got a rude shock. There were three of us reporting aboard that day, and I learned that they needed only two for the submarine DirSup program for the next two years, but they were undermanned in the airborne program.

My new boss noted that I was also a volunteer for flight operations, the only one of the three new arrivals, and he called me in to discuss this. He explained that if I did not voluntarily switch to flight duty with the fleet air reconnaissance squadron (VQ-1), the three of us, the newly assigned officers, would need to draw straws. One of us would have to be assigned to a desk job for the next two years.

So, if I did not honor my "volunteer to fly" statement, there was a good chance one of the three of us, myself or one of my classmates for the last 22 months, would be completely screwed. The plus in the deal was that I would have to go back to the US for two more weeks of Survival, Evasion Resistance, and Escape and Deep Water Environment Survival Training (SERE/DWEST) instruction in the San Diego area.

These two courses were required of everyone who flew in hostile airspace. They were the final requirement to qualify me to fly as a "special evaluator". The WLR-6 course in Groton and the OinC courses at NSA gave me all the other qualifications I needed. The real plus was getting to go back to San Diego. I loved old San Diego. It was a sailor's town in those days, and I had enjoyed my time there on HORNE.

I thought it over for a few seconds, not all that many, and then said: "OK, I'll go fly." I was an authentic dual volunteer. I wanted to do both, and I immediately realized that I would generate some goodwill if, even though I was highly disappointed, I cheerfully agreed to change my career plans and switch from what I had just spent nearly two years training for and went flying.

One must understand that I had just been brainwashed for the past 22 months that the submarine reconnaissance program was the absolute pinnacle of the US's intelligence collection program. Not only the Navy. Not just of the armed forces but the entire US intelligence community. The books "Blind Man's Bluff," "Stalking the Red Bear," and "Red November," which were published in 1998, 2009, and 2010, respectively, make that point in spades.

So, I was giving up, or at least delaying, a significant opportunity. My new boss did assure me that they would bring me back into submarine operations at the earliest opportunity. Still, as he honestly pointed out, it would probably be two years before they would need me, and there were no guarantees they ever would. I thought I had two choices, get screwed with a smile on my face or get screwed. As it turned out, this seeming screwing turned out to be yet another massive blessing in disguise, but it took years to recognize it.

Two weeks later, I was on an airplane bound for San Diego and SERE/ DWEST training, leaving my wife and 1-year-old son to set up our new home on the economy in rural Japan. Living off-base was a lot rougher than living on-base due to the lack of stores and shops nearby. Still, it did have its rewards. We subsequently made many friends both among the other US forces personnel living around us in an American 'ghetto,' and among the rural Japanese, who were eager to meet and befriend all Americans.

SEALS go through the same training near the end of their curriculum, but for them, it is a walk in the park at that point. It is more than a bit challenging for the rest of us. However, on the recommendations of others who had gone to SERE/DWEST training, I made a point of physically training up during the short time I had. SERE trains you on how to survive, evade in enemy territory, resist interrogation, and survive in a prison camp, as well as how to escape if given a chance. It reminded me of some of the outdoor games I had played at the various scout camps I had attended, so I did have some advantages.

However, one guard, who looked like he was moonlighting from his regular job as a National Football League defensive lineman, did not like my answer to his question. He grabbed the front of my jacket, easily lifting my 185 pounds off the ground with one hand. In his other hand was a Colt .45 revolver, commonly known as a "hog leg" due to its size, the barrel of which he jammed up my nose, seriously cutting the inside of it with the blade sight.

He repeated his question, and I remembered what they had told me to do when it appeared you were in mortal danger. I was not sure I was not, even though this was supposed to be just a training situation, so I gave him the answer that I thought he wanted, and he dropped me to the ground like a rag doll. He looked a lot like Rosy Grier, an NFL All-Pro lineman of that era, except this guy may have been even bigger, and I have wondered to this day who he was. Maybe he WAS Rosy!

I understand the prison camp guards went to a special school about how to scare us without seriously hurting us. It must have been one hell of a school. The other thing I very clearly remember was being "water-boarded" several times. I thought I might die on the waterboard accidentally. Others tell me that they had the same thought. It was definitely not fun, but you do learn a lot, especially about yourself, precisely what the objective is, I now realize.

DWEST teaches you, as its name implies, how to survive if you find yourself adrift on the sea. The part I most clearly remember was being dropped from the back of an underway utility boat rigged with what appears to be a swing set, except the seats are parachute harnesses. You put on the harness and then are dropped into the sea as the boat travels at about 6 knots. This simulates being dragged by a parachute that has failed to collapse when you land in the water.

WLR-1 ELINT RECEIVER
My first piece of SIGINT gear

In my case, the harness partially separated shortly after I entered the water, and instead of being dragged heads up, I was being carried upside down, with my feet above the water and my head well underwater. I realized what the problem was and was trying to pull myself up the simulated parachute riser to get my head up when one of the safety observers noticed I had a problem and leaped in and saved me. The chief instructor was sure I had screwed up somehow and panicked, which I most definitely had not.

When I showed him the separated harness, he pulled out his knife and cut it into pieces. He asked if I wanted to do it again, and I readily said yes, so they hooked me up and dropped me once again, this time with no problem. It was fun! But it does go to show how dangerous it is to ditch into the sea. I finished SERE/DWEST and headed back to Misawa via Yokota Air Base. Things were about to get most interesting, but I had no idea what was about to happen at that time.

Shrike Hit!

Chapter Four

As I walked into the passenger terminal at Yokota Air Base, I was surprised to hear my name being called over the public address system. I had just arrived by bus after a many-hour flight from San Diego to Haneda International Airport. Tokyo's only international airport in those days. Haneda is on the far east side of Tokyo, and Yokota is well out to the west. The bus was a USAF school bus, so it had been yet another long military "no-frills" bus ride. This time across Tokyo, a vast city. That trip was on top of a long, full flight in American Airlines Economy Class across the Pacific from San Diego. So, I was dog-tired by the time I walked into the terminal at Yokota Air Base.

I had just completed the two weeks of instruction in survival, evasion, resistance, and escape/deepwater environment survival (SERE/DWEST). These courses completed my qualifications to fly in Fleet Air Reconnaissance Squadron ONE (VQ-1) aircraft, the Navy's "Spy Squadron" in the Pacific, in hostile airspace. I had no idea why they were calling me. But, the message was clear, "Lieutenant Junior Grade George Guy Thomas, Report to Navy ATCO (Assistance Company) immediately." I had little idea that answering that page would be the start of the most memorable three months of my Navy career... Indeed, probably of my life.

The ATCO had new orders for me. They directed me not to return to my family in Misawa, Japan, 600 miles north of Tokyo, but rather to board the bus they had waiting for me. I was to proceed to the USS WORDEN (DLG-18) immediately. She was a guided-missile cruiser in port at Yokosuka Naval Base, nearly 30 miles to the southeast, but still in greater Tokyo, to assume duties as the assistant officer-in-charge of its Naval Security Group (NavSecGru) DirSup Detachment for up to 2 months.

They also provided a phone number to the ship and a point of contact, Lieutenant (LT) Hugh Doherty, a fellow NavSecGru officer and a neighbor at Misawa Air Force Base. Before I left Yokota, I was able to get through to him. He told me WORDEN was assigned to conduct a late-breaking "special operation," we might be gone for as long as two months, but probably shorter. He would tell me more when I got on board.

Hugh and I had trained to run technical intelligence collection operations, so I knew without being told we could not discuss the details of the operation nor our role in it on that unclear phone circuit. I would learn the details when I reported on board and got to a space (maritime for ROOM) in the ship cleared for TOP SECRET. In that I had just spent nearly two years training for this type of mission, it was not unwelcome news.

I was a bit distressed to be leaving my wife to care for our 20-month-old child in rural Japan, but Hugh assured me all was well and that the other wives of the Direct Support (DirSup) Division would be looking out for them. They had a lot of experience in this as all their husbands disappeared for two or more months to run the Navy's technical collection operations in the northwestern Pacific and adjacent seas in "direct support" of naval operations, hence the division's name.

DirSup had been an integral part of naval operations almost since the advent of radio, really coming into its own during World War II. They were an integral part of the Navy's highly effective submarine campaign against the Empire of Japan and in air defense operations defending the task forces supporting the landings at the various Japanese held islands. Still, its operations remained highly classified and never discussed outside of appropriately cleared "spaces." I was immensely proud to be joining this tradition of unknown excellence, the "Silent Warriors" of the Navy. I grabbed my seabag and jumped on the

NAVY shuttle bus, headed to Yokosuka. It was a three-hour ride, and I was already exhausted, but I was so excited to finally get the chance to employ my training that I did not sleep a wink.

Three and a half hours later, I was walking up the gangplank of USS WORDEN. It was my first ship assignment both as a naval officer and as a NavSecGru DirSup officer. I had been on a very similar ship for 20 months as Yeoman (Navy for clerk) assigned to intelligence duties. Indeed, for the last five months of the tour on USS HORNE, I had been the de facto Intelligence Officer on the ship. That was what got me selected for the elite NavSecGru, but this was my first time to walk up a ship's gangplank for duty as an officer. I was exhilarated but a bit apprehensive.

I found Hugh's bunking space and joined him in the wardroom for a cup of coffee and to get a quick brief at the unclassified level on what was up. He apologized for my being diverted from going home, but I was not angry at all. This was precisely what I had trained for, and I was anxious to get underway on my first assignment. I learned we were going to the Vladivostok operations area (OpArea), home of the Soviet's Pacific Fleet, to "demonstrate freedom of the seas" and collect intelligence on the Soviet Navy and any Soviet air units that reacted to us. We had intelligence that the Soviet Pacific Fleet was preparing for an operational exercise. Our job was to provide early warning of any evidence of hostile intent and run the intelligence collection operation. Even today, I cannot discuss in detail what our priority targets were, but the White House had cleared the mission.

The whole operation had generated very quickly, and we were on board for a very specific purpose. We spent the next two days unpacking our technical intelligence collection equipment and checking it out. This equipment was so classified and expensive that it was not left on warships full time but rather taken on the ships as they prepared for "special operations" such as this one.

On our last night in port Hugh and I had a big dinner with a nice bottle of wine at the Yokosuka Officers' Club and came back to the ship about 9 PM to get to bed early as we were getting underway early the following day. We were to get the ship's mission brief and be introduced to the wardroom, the ship's officers, as soon as we cleared Tokyo Bay.

The crew had seen us as we came on board with our equipment, but they were told not to talk to us or to even speculate out loud as to why we were onboard. That was standard procedure for the DirSup Spooks in those days. This may still be, but maybe not, as Signals Intelligence (SIGINT) collection is now openly acknowledged, though the details remain highly classified. There is a great deal more in the press these days about NSA and signals intelligence (SIGINT), our primary task.

As we arrived back on board that night, we noticed a lot of talk on the ship about the fact that North Vietnam had sent its regular army into South Vietnam in a full-scale invasion. Still, while we noted the news with professional interest, we basically ignored it. We were going north to the Russian coast, not south to the Gulf of Tonkin, right?

All that changed the next morning when the WORDEN's executive officer, the 2nd in command, briefed the wardroom, the Navy term for all officers of a ship, that the main battle army of North Vietnam had invaded South Vietnam the day before. It was possible the war was going to change dramatically, and we may be diverted to assist that effort. Until that moment, Hugh and I had naively not thought about it much as we "knew" we were going to the Russian coast.

Our inexperience in the ways of the Navy deployed in the Western Pacific was showing. An hour later, as we cleared the Yokosuka sea buoy marking the naval base entrance, and headed out into Tokyo Bay, CAPT George Schick, WORDEN's commanding officer, came on the 1MC, the ship's generally announcing system. He informed the crew that our orders had indeed changed. We were turning south, not north, and proceeding to the northern Gulf of Tonkin at best speed to assist in the most extensive build-up of naval forces since the landing at Inchon in the Korean War nearly 20 years before. We were to assist in the aerial and naval bombardment of North Vietnam and its army in South Vietnam.

Hugh and I looked at each other in surprise and shock. We had been shanghaied! We had optimized our expensive technical intelligence collection equipment for Sea of Japan targets; thus, much of it would be useless in the Gulf of Tonkin. Also, our team consisted of area specialists for the Northwest Pacific, not the Gulf of Tonkin. Indeed, apart from one E-8 marine communications specialist, I was the only member of our team who had been to Vietnam.

USS WORDEN (DLG-18)

During my short time onboard WORDEN, I learned that she was understaffed with junior officers. They needed qualified combat information center (CIC) watch officers and officers-of-the-deck. That shortage gave me an idea. In that I had spent nearly two years on a very similar ship, USS HORNE (DLG-30), much of it on the bridge or fantail as a phone talker, I was very familiar with the duties of officers on underway watch. I had also spent almost six months as the acting intelligence officer, standing watches port & starboard watched in CIC. 12 hours on, 12 hours off, but on-call 24/7 and while off the coast of North Vietnam it was more like 18 hours on, six off, if that much, so I was familiar with the equipment in CIC. Even though a junior enlisted man on HORNE, I also studied how to use every piece of CIC equipment, especially the WLR-1, ULQ-6, and ALR-91 electronic warfare equipment, which fascinated me. I had also learned how to use all the radars (air search, surface search, fire control) and the computer systems running the Navy Tactical Data System. I was no expert, but I did fully understand them.

As I have mentioned earlier in this book, I was also interested in a career at sea early on, and at age 15, I had joined the Sea Scouts. Both on HORNE and before that as a Sea Scout, I studied how to be an Officer of the Deck (OOD), hoping to be one someday, a goal that I had put on the back burner as I learned how to be a "spook." But now I saw a chance, and I had no doubt I could, with my background, assist the ship both on the bridge and in CIC. It certainly beat sitting in the officers' mess drinking coffee day in/day out as our intelligence collection spaces were much too small for just sitting around. Think of six men sitting back-to-back in an econovan full of electronic gear with inadequate air conditioning.

I thought about it for the morning, and I knew what I wanted to do by lunch. I went to Hugh and asked his permission to volunteer to stand ship's watches. He was doubtful they would let me do it. But he had no real idea how much direct, pertinent experience I already had, even though I was just Lieutenant, Junior Grade (LTJG). He planned to spend his time training the team and himself to become better systems operators and analysts. Still, there was little room anywhere on the ship to discuss highly classified matters, and technical intelligence was even more highly classified then than it is today. He was probably happy I had found something productive to do, and I could train with the team when I was not on watch.

I sought out the senior watch officer, a lieutenant commander (LCDR), and offered my assistance. He was the officer responsible for manning the bridge and CIC with qualified officers and was, as Hugh had predicted, a bit skeptical. I briefly laid out my experience and mentioned my interest in a naval career. In those days, many junior officers were in the Navy to avoid the Draft, but I assured him I was not one of those. We both knew that qualifying as an OOD could not hurt; indeed, it might even be a "tie-breaker" for a promotion sometime in the future, and besides, I was genuinely interested. We went to CIC, and he asked me to describe the purpose of everything I saw there, which I did.

He remarked he wished his other junior officers were as checked out as I was. And we went to the bridge, where again he tested my knowledge. I passed easily. He was glad to have another qualified officer on his watch list. We both knew I needed to be trained in how WORDEN did things, so he quickly outlined a three-day training program to bring me up to speed. That was the standard procedure for new officers with prior experience assigned to any ship.

My first watch under instruction was in CIC, and throughout that first watch, I demonstrated to the watch team that I knew the systems and how to use them, cold. I had not wasted my time on HORNE, even if I was just a junior petty officer. My following few watches were shifted to the bridge, where I did need a bit more training but still qualified as a Junior Officer of the Deck in two days. The LCDR senior watch officer said that was record time for a junior officer in his experience (I did have a running start!). He scheduled me to stand my first full watch not under instruction commencing 0345 16 April 1972.

It was a fateful time to be assuming the watch as many miles away, in Washington DC, the decision had been made to commence using B-52s against the Haiphong, North Vietnam's major port, at 0355 that day, 10 minutes after I assumed the watch.

At SERE/DWEST, I had not shaved for several days and had liked the red fuzz that sprung up. I almost kept that young beard, but I had decided to shave it off. However, as I dressed for my first at sea watch as a naval officer, I again considered growing a beard and decided for it this time. That decision got me on the bridge a few minutes early. That timing proved critical and may well have saved my life.

Before assuming the watch on the bridge, all officers are required to visit CIC to get a situation update. I dutifully visited CIC as I had been trained to do, but the situation there was anything but routine. I had not been told of the impending B-52 strike the night before and did not know we would be as close to Haiphong harbor as we already were, and we were moving closer. I learned we had moved well north of our previous position in the north-central portion of the Gulf of Tonkin and were now very close to Haiphong, North Vietnam's major port. WORDEN was going north to position its Search and Rescue (SAR) helicopter, as close to the action as possible. It had limited range and could only hold 2 or 3 passengers due to its being heavily armored. It might need to make multiple short runs to pick up the eight-man crew of any B-52 shot down over the harbor. My first watch promised to be anything but routine. We were all excited to be participating in history in the making and would probably have a ringside seat to the real action…Little did we know…

One other brief in CIC was of particular significance. Electronic intelligence (ELINT) had detected a brief intercept of the radar associated with North Vietnamese PT boats. It could also be the radar on a Chinese or Korean submarine. Still, none were known to be in the area, and it was deemed highly unlikely a submarine would be operating in this area. If it were, it was even more unlikely it would be employing its radar. Still, it did make us think, and the several watch teams, sonar, radar, ASW torpedo, guns, and missile were all at full alert but still not fully manned.

This was also true of our damage control teams. We were in full EMCON, not radiating anything, which was usual when you are that close to a hostile coast and did not want to be detected. Once you are discovered, there is little reason not to radiate, but we were hoping we had not been detected at that time. We were watching the tactical situation over NTDS. Our radar screens were updated by other ships in the task force as their information was data-linked to us.

After getting the tactical situation in CIC, I went to the bridge, which was just forward on the same level on this class of ship. I reported to the Officer of the Deck (OOD) that I was ready to commence turnover, and the on-watch JOOD, a chief, and I began our turnover. The first thing JOODs routinely do on turnover is inventory the classified material on the bridge, including the tactical codebooks. These books are kept in a strong storage cabinet, serving as a safe located in the bridge's right-center. After we had jointly verified all classified materials were "present or accounted for," we joined the on-watch OOD and his relief for a brief discussion of the tactical situation. We could clearly see Haiphong. At least one fuel tank was on fire from an earlier fighter bomber attack.

Shortly after we started our conference in the bridge's front center, we received a report from the aft lookout on the fantastic that electrified us all.

"Bridge, Aft Lookout. Two jet aircraft just passed low over us, lit their afterburners, and have commenced a climbing turn off our starboard quarter. I think they may be coming back to attack us. I am going off-line to seek shelter." A few seconds after his warning, I saw two flashes out the bridge's starboard door. A second or two after that, I heard what sounded to me like two 5"/38 naval cannon rounds fired in quick succession. I happened to be the left most of our group of four, facing the ship's starboard side, so I knew the flashes had come from somewhere aft of us on the starboard side.

Also, the 5"/38 naval gun is the main battery (weapon) on the destroyer accompanying us. It was located aft of us on the starboard side, so I assumed it was firing at either the hostile aircraft or surface targets, probably the NV PT boat (or boats) reported to be in the area by ELINT. I immediately guessed, incorrectly, that we were under a combined attack of aircraft and patrol boats, but I was not scared. I was definitely not aware we were under air attack. The fact that the aircraft had afterburners meant that they were either US aircraft or, if North Vietnamese, were MiG-21 fighter aircraft with no ship attack capability, so there was no air threat. Funny the things you think of in combat and how fast your mind works. Also, funny how wrong you can be!

I took a step or two toward the starboard wing of the bridge, thinking I might be able to see the PT boat, even though it was a very dark, moonless night. Back in those days, my night vision was famous. I could see many things no one else could see at night. I knew that was true when I was a Boy Scout but had not thought about it until I was on HORNE.

As the aft lookout on HORNE, I would report things no one else could see except through the night vision glasses. It got to be a game, and my fellow crewmen on HORNE used to test me all the time. I demonstrated time and time again that my eyesight, both day and night, was very unusual. I had not fully understood until I went to sea in the Navy. Thus, it was not unrealistic that I hoped to help fight WORDEN by using the old mark one, mode zero eyeballs. I also knew boats moving at high speed made both a bow wave and a wake which was often very visible, at least to me, at night.

All my thoughts of looking for signs of PT boats changed literally in a flash and a thunderous explosion high up on our mainmast. The ship shook, or more accurately, shuddered, but not too violently. Again, I was not too scared. (I guess I am a slow learner.)

My immediate thought was something like: "Those PT boats hit us with a small missile, maybe an anti-tank weapon of some type...Lucky shot!" I took another step or two toward the starboard bridge wing.

Those few steps may well have saved my life as the whole bridge, especially the overhead (ceiling to landlubbers), exploded with a very, very loud "BANG." It was much louder than the just previous explosion. There were pieces of red-orange, extremely hot shrapnel flying all over the place, ricocheting off everything: the floor, the radar scope housing, the charts table, the walls, the lee helm with its engine-order handles, the armor plate windows. The only place where the shrapnel was not thick seemed to be right where I was. It was swarming all around me, and I now know the other three most senior people on the bridge. But the shrapnel had hit every other person on the bridge, seriously wounding almost all of them. The explosion knocked all of us down, and at the time, I thought I was the only one not seriously hit. I did take a piece of shrapnel through the thumb, but I was uncut other than that.

Everyone I could see was bleeding profusely. However, the concussion of the blast had blown me into the air. It was like flying through the middle of a big 4th of July sparkler, and I had received a stiff blow to the head when I bounced off the deck on landing.

As I was airborne, I saw, out of the corner of my eye, the antisubmarine rocket (ASROC) launcher, an eight-cell "pillbox launcher" full of rockets on the foredeck just forward of the bridge, taking what I thought at the time were multiple hits.

In retrospect, they were probably burning debris from the first missile that had hit our primary air search radar a few seconds before, landing on it. But, I did not think of that at the time.

In a combat zone, the ASROC launcher was fully armed with eight missiles. Many of the "sparkler" hits bounced off the launcher, but others clearly seemed to penetrate it. Luckily, none hit and ignited either the warheads or the rocket fuel. Still, the smoke trickling from the launcher (from the debris?) seemed to indicate something was on fire in the launcher, which I knew to be fully loaded with live rocket rounds with powerful warheads designed to sink submarines. Not good! Not good at all! Now I was scared!

**F-105G Wild Weasel
with both Shrike and Standard Missiles**

The concussion had knocked most of us a foot or more into the air. The picture of the damage to the launcher in my mind to this day is from an angle that you cannot see from standing on the bridge unless you are very tall or on a ladder. I clearly saw the top of the launcher getting hit. I knew the pillbox launcher was fully loaded with live ordnance, and rocket fuel ignition was of particular concern. I could easily envision the fuel fire igniting a warhead, which would probably have ignited the warhead next to it, starting a chain of explosions.

That launcher was right on top of the Terrier missile magazine, the cruiser's primary weapon system. It, too, was full of fully armed and fueled long-range missiles with lots of fuel and smaller but still deadly warheads. If they exploded, the front of the ship could be blown off. It occurred to me that I probably had just a few more seconds to live. (A healthy imagination is not a particularly good thing to have in combat...)

I instantly decided that I was probably the only surviving officer on the bridge. I knew that we had been hit by two small, short-range missiles, probably anti-tank rounds, and I believed North Vietnamese PT boats were launching them at us. Thus, I thought the best thing I could do would be to ensure our lights were off and to set General Quarters. That would make it harder for the PT boats to see us and accurately launch torpedoes at us, still a major concern of mine. It would also get the damage control crews up and on their way to their stations before the bow blew off, killing everyone on the bridge or taking torpedo amidships. It is funny how your mind works when faced with probable death.

My one thought was to try to save as many of my new shipmates as possible. I called out "Sound General Quarters!" and heard the Quartermaster of the Watch (QMOW) respond with "Sound General Quarters, Aye, Sir!" I did not have the authority to issue that command, but I was not sure who was still alive on the bridge. It seemed entirely possible to me that I was the senior surviving officer on the bridge. I subsequently discovered this was not true as I heard a familiar voice call out, "What course to steer?"

I was still more than a little dazed from the concussion of my head bouncing off the deck at the end of my flight through the air. My dazed state became apparent when I tried to remember the reciprocal of northwest (the direction of Haiphong) as digits, so I called out "180," which would get us out of the immediate area of Haiphong, which was north-north-west of us. I knew we would need to turn about 40 degrees to port off 180, but I could not remember if that was 140 degrees (the right course) or 220, a terrible course. Course 220 would take us further into danger by heading us straight at Haiphong and the North Vietnamese coast.

About this same time, our port 3'50" gun battery opened fire, which further heightened the impression we were under PT boat attack. I also knew we had slowed the ship not long before the hit and knew we needed to increase our speed.

The off-going OOD had regained his feet and alertly grabbed the wheel, which was running free because the helmsman was seriously wounded. I discovered that he was the person I was talking to when I staggered over to see who had the wheel. He had the wheel, but there was no one on the engine order telegraph, which sends commands to the engine room to tell it to set a specific speed. That person had been hit, too. There are two indicators on the engine-order telegraph: One reads speeds in the old traditional way, from "Ahead Flank" through "ALL Stop" to "Back Full." The other is a digital display with three numbers. "0-0-0" to "9-9-9" You are supposed to move the handles to the desired speed range "FLANK" "FULL," "HALF," "SLOW," and then fine-tune it with the digital display. The speed range indicator was no problem in understanding what to set. We needed "FLANK." But we also needed to select the appropriate digits. What was the setting for our top speed? In my dazed state, I could not remember that either. Finally, either the OD or I, I really cannot remember who, rang up "FLANK," and I dialed in 9-9-9 (three times our actual top speed of 3-3-3). I figured the guys in the engine room, who surely would have heard the two explosions and the call to GQ and would get the message we wanted everything they had. They did, and we rapidly built up to our maximum speed.

In talking with the engineers later, we all had a bit of a rueful laugh. They understood completely. They, too, had heard there was a PT boat in the area and were more than happy to give us every knot they could squeeze out of her. They were not anxious for us to take a torpedo, especially in the engine room, where they worked. It would have been an instant, very unpleasant death for them as the scaling oil and water exploded in their very crowded spaces.

Shortly after that, I realized some of our running lights were still on full brightness, and I walked over to the panel that controlled the lights. Again, I was too dazed and flash-blinded to figure out how to turn the lights off. I was standing there trying to figure out what to do when the Quartermaster of the Watch (QMOW) came from the navigation "shack" just behind the bridge and asked what I needed. I told him we needed to turn off our lights as there was probably an enemy PT boat stalking us. Enough said!

He quickly reached up and expertly flipped the master switch turning off all external lights. I congratulated him and turned my attention to helping the many wounded. I was trying to stop their bleeding any way I could, primarily by making bandages from their clothes. Wanting to get the wounded off the bridge, I asked the QMOW to pass the word for the corpsman to lay (go) to the passageway outside of the Captain's in-port cabin.

I remembered on HORNE that the cabin was designed to be easily converted into an emergency room. And it was one deck immediately below the bridge and thus accessible to get wounded too, either by carrying the most wounded or, if they were able, on their own. This action did have the unfortunate effect of suggesting to many of the crew that the Captain had been hit. Not my intention at all.... an unintended consequence, for sure.

As I worked on stopping the bleeding from seven men, another minimally wounded person came up and reported at least one seriously injured person on the port bridge wing. I was not happy to go out on the weather deck, which we all thought we might be raked by machine gun and small cannon fire at any moment.

Still, I did not want to leave my shipmate out there, either, so off I sprinted. "Get him and get back" was running through my mind. I quickly found him lying in a pool of blood. In looking for his wounds, I found the back of his head was crushed. The concussion that blew those of us on the bridge into the air must have thrown him into the bulkhead. I cut a cradle for his head from his shirt and carried him down to the passageway I had just asked the corpsman to go to outside the Captain's cabin. When I got down to the corpsman, he was already hard at work professionally bandaging the wounded I had sent down to him with my makeshift bandages.

I carried the young man into the room and asked the corpsman to look at him immediately as he was very seriously wounded. The corpsman immediately moved over to the man in my arms and conducted a quick examination. He looked up at me without saying a word. I will never forget the look of hopeless horror on his face. There was nothing anyone could do except make the young man more comfortable in his last moments. I went back to the bridge and saw the situation was pretty much returning to as normal as possible.

One of the other junior officers saw me and reacted with horror. I had to ask him, "Why are you so horrified to see me?" and he asked.

"Are you wounded?"

"No, but I have been helping the wounded." "Have you seen yourself in a mirror?"

"No, why?"

"Go look!"

Since my job was pretty much done on the bridge, I went down to my cabin and looked in the mirror. I was completely covered and soaked in blood. I peeled off my khakis and threw them into the shower, climbing in after them. I was so exhausted and emotionally spent I sat down in the shower and nearly went to sleep there…. With the hot water running on me.

Later the next day, CAPT Schick made a point of seeking me out and thanking me. I felt wholly inadequate, a man had died basically in my arms, and I did not know what to say. So, I just said, "Thank you. I am sorry a man died." He also showed me a piece of one of the missiles that had hit us. It was only a couple of inches square, but the marking was clear – "Made in the USA." The fragments of the two missiles that hit us were all over the ship's topside. They told the story that two US-made SHRIKE anti-radiation missiles had hit us. I felt sick.

I learned much later that the SHRIKE anti-radiation missile was indeed made from the 5"/38 shell, so I was not entirely crazy in thinking I was hearing a 5"/38 firing two rounds. It is just that I never considered they were coming from an aircraft and were aimed at us.

Why us? Once I knew it was a SHRIKE that had hit us, and we had been radiating our SPS-48 air search radar briefly just before the missiles hit, the rest was easy to figure out. That radar operates at the same frequency as the FAN SONG B, the missile target acquisition radar of the Russian surface-to-air missile system NATO calls the "Guideline." It was widely deployed in North Vietnam.

Both the Navy and the Air Force flew aircraft equipped with SHRIKEs whose mission was to suppress enemy air defenses (SEAD). The seekers in those missiles were designed to home on that frequency. We were close enough to the shore that two of those aircraft had mistaken us to be an enemy surface to air (SAM) site. Detailed investigation of the damage done by those two small missiles was revealing. We lost our primary air search radar, the SPS-48, three of the four missile fire control radars, the SPG-55, our secondary air search radar, the SPS-43, and one of our two surface search/navigation radars. Judging by the shrapnel impact patterns on the upper weather decks, it was easy to see most of the damage to our radars was done by the first of the two missiles that hit us. The second missile had hit the roof of the bridge, right over where I was standing. Luckily for me, the missile's warhead was designed to spray very lethal small steel cubes out to the sides but not straight ahead. I would have died right there had it done so. The SHRIKE was a highly effective weapon.

CODA

1979 – Osan AFB – I subsequently served as a technical intelligence officer in special mission submarines. I also became a systems operator on Navy reconnaissance aircraft and did an exchange tour with the USAF. That tour is worth its own story and is covered in a later chapter, but that is not essential to the SHRIKE Hit story except in one regard. While on a USAF pre-Inspector General visit to Osan AFB, I met Colonel (Col) Dan Berry. He had orders to the Electronic Security Command, San Antonio, where I was stationed.

He wanted to talk to me about what it was like there and get an idea of what to expect. During our discussion over a beer at the Officer's Club, he mentioned he flew with the Wild Weasels in Vietnam.

That was the squadron that I understood had hit the WORDEN, so I told him my story. He assured me it was not his squadron. We compared a few more notes, and he was not assigned to the squadron until about two weeks after the incident. Still, he was sure it was not the Wild Weasels. I was an LCDR, two ranks junior to the Colonel, and did not wish to argue with a senior officer, especially since I had no actual proof, so I let it drop. Indeed, we enjoyed an excellent professional relationship when he did check in to San Antonio. He was a fine officer and gentleman.

In due time, I transferred to the Naval War College and, when I had completed the course of instruction, I was honored to be asked to stay on as research staff. I was assigned to the Center for War Gaming in the Center for Naval Warfare Studies, the research side of the NWC. Our research took us to many places, including the Pentagon.

One day I was walking down a hall there when I ran into Colonel Berry. We greeted each other like long-lost friends. He was up from San Antonio on Air Force business, but the first thing he said was. "I was talking about you just last night. Had dinner with my Wild Weasel squadron commander. First time I have seen him since Vietnam. Told him your story. Do you know what he said to me?"

"Colonel, I have no idea!"

"His first words were, 'I never told you about that?' I owe you an apology, Guy!" Your story was true and accurate!!"

"Colonel, I never doubted that!"

Several years later, one of the two USAF pilots that hit the WORDEN, with another pilot who flew another Wild Weasel that night, made a point of looking me up and taking me to lunch, where we spent a couple of hours swapping war stories. They were most sorry to have hit us but were gratified to learn from someone who had been on the receiving end as to how much damage one Shrike could inflict.

To be fair to the Wild Weasels, they had probably the most challenging, most dangerous airborne job in the war. Their motto, "First in, Last out," gives a hint, but only an indication. Their job was to get the enemy air defenses to light off their electronics and go through their missile launch sequence. The Weasels had sensors onboard that detected that sequence, and they tried to destroy the missiles, their launchers, and the controlling radars before the cycle could be completed and the missiles launched.

Often the missiles were launched and the Weasels had to dodge them as they bore down on the launch site to fire their SHRIKEs. They then used the exploding missiles' flashes to pinpoint other high-priority targets in the area, such as the command van and unlaunched missiles to follow up with cluster bombs.

In my specific case, the pilots who launched the Shrikes that hit WORDEN saw from the flashes of the two exploding missiles that they had hit a cruiser and, knowing that only the US and possibly the Russians had cruisers in the area, had aborted the bombing run. Had they followed through with that bomb run, I probably would not be alive to be writing this now.

The Weasels were truly dueling with death and, while they won that duel many times, many of them did not come back. Not sure which group had the highest mortality rate, but the Wild Weasels were among the highest, if not the highest. The guys who flew post-strike reconnaissance in the Navy RA-5C Vigilante and RF-8G Crusaders, and the Air Force RF-101 Voodoo must also be running as the most dangerous mission over North Vietnam. They all have my utmost respect. As the recce guys say, "ALONE, UNARMED, AND UNAFRAID... *Hey, two out of three is not bad...*"

****

For an excellent history of the Wild Weasels during the Vietnam War, see "The Hunter Killers" by Dan Hampton. It is an extraordinary story of what it was like to be a weasel and the "weasel mission." The story is gripping in many places. Indeed, I could "feel the g's!" as the author describes several of the epic weasel versus SAM battles. And I grieved when an American airman was lost. However, the book does not mention the attack on the USS WORDEN. It is also an excellent overview of the political and military history of the 20 plus year lead-up to the war and then the political side of the war itself and China and Russia's roles in the war. Highly recommend it!

MINES, MISSILES, and MIGS

APRIL-JUNE 1972

Chapter Five

The two SHRIKE missile hits on the USS Worden were a "mission kill" on her. All but one of its radars were out of action, and we headed for Subic Bay in the Philippines at our best speed, which was something over 30 knots.

Along with the rest of the Naval Security Group (NavSecGru or NSG) team, we spent the transit to Subic Bay preparing our technical materials, which were highly classified, for shipment via the Armed Forces Courier Service (ARFCOS) back to our base at Misawa, Japan. Shipping highly classified material was a task you do not take lightly. A screw-up could cost you your career.

Upon arrival at Subic Bay Naval Station, we went straight to the ARFCOS office to arrange their shipment back to Japan. At the ARFCOS office, we were handed a message from the ranking NavSecGru officer in the area, CDR Paul Deschler, Commanding Officer, Naval Security Group Activity San Miguel. It was located at the Naval Communications Station on Lingayen Gulf, about 20 north of Subic Bay. He knew of our problems, and both offered us all the support his command could provide as well as an intriguing, at least to me, opportunity.

The buildup of naval units to support the increased operations against North Vietnam had put a real strain on NavSecGru Activity San Miguel. The need for NSG Direct Support teams to crew the many new ships in the area to provide technical intelligence support had well out-paced San Miguel's ability to supply them. Knowing that, the Pacific Fleet

NavSecGru commander had given them permission to use our detachment from WORDEN any way he needed to assist the increased war effort.

    CDR Deschler did not want to keep anyone against their will, but he offered us the chance to volunteer to become part of his direct support teams deploying in the Gulf of Tonkin to assist the war directly. Most of the guys on my team had focused on the Russian problem for most, if not all, of their careers. They did not feel they could do a credible job without significant retraining. On the other hand, I had already served in an intelligence role in a missile cruiser off Vietnam, and I was sure I could come up to speed quickly, so I volunteered. An old Army adage says,

    "Marching in the direction of gunfire is never a bad idea."
So, it proved for me.

    Another old military adage worldwide is "Never volunteer for anything." But I am convinced that is dead wrong, and that has certainly proved accurate several times in my case. I enrolled in a quick course on the North Vietnamese order of battle, capabilities, and tactics. I was also given training on the various gear my team would be using and the communications procedures we would be following. Nearly all of it was familiar to me due to my time on USS HORNE, and I quickly mastered what else I needed to know. During this short period, the officer in charge of DirSup introduced me to my team leader, an exceptional young man, Ensign Gary Blank. He was very bright and personable, one of the sharpest men with whom I ever served. Even though I became a full Lieutenant (LT) on 1 May, which put me two ranks senior to him, I had no problem accepting him as my leader.

    I completed my indoctrination and training very quickly, and our direct support team embarked on the USS CHICAGO (CG-11) in port Subic Bay in early May. The maintenance support team in Subic had already checked out our collection and communications gear, and we sailed for Yankee Station in the northern Gulf of Tonkin the next day.

I spent both the transit and our first few days on station getting to know the current mission, our equipment, and our men's capabilities. We had four intercept operators, two communicators, plus a couple of maintenance men to keep the intercept and the special communications gear online at all times.

Gary Blank and I got on very well, and I relieved him early on the morning of 8 May 1972. He immediately departed CHICAGO on the regularly scheduled logistic helicopter. He was en route to the USS OKLAHOMA CITY, another WWII-era cruiser converted to a missile cruiser. It was the fleet flagship. And he assumed duties as the signals intelligence advisor to the task force commander.

I was now the Officer in Charge of my first signals intelligence collection and reporting detachment.

Almost immediately after I assumed my new role, the ship's operations officer called a short briefing in the wardroom for all officers. I learned we were on our way north to support the mining of Haiphong harbor by Naval aircraft.

The various military commanders had long requested this action, but the politicians had not wanted to escalate to this point. The North Vietnamese all-out invasion of South Vietnam a month earlier had changed all that. President Nixon and his national security advisor, Henry Kissinger, had decided it now had to be done. We were to take a station close to the harbor's mouth, right back in the same spot where the WORDEN had been hit three weeks earlier.

We all quickly dispersed to our various duty stations. I hurried to Supplemental Radio (SupRad) as our intercept spaces were named. Once there, I briefed the team on the news, and they all cheered. My team and I knew we needed to be exceptionally watchful this close to the hostile shore. Indeed, we soon could hear the North Vietnamese' shore batteries at the mouth of the harbor firing at us. We could even feel the concussion of the shells hitting the water all around us. It was not accurate fire because we were jamming their fire control radars, but it still felt very, very eerie. It helped to sharpen our senses as to what was happening on the airways.

A short time later, we could also hear and feel the mining aircraft, Navy and Marine A-6s, and A-7s, as they screamed in at very low-level right over the top of us. The men on the ship's bridge had a front-row seat as the planes roared in on their targets, pitching up to loft-launch their mines into the entrance of the harbor just after they passed over us. After weapons release, flight physics required them to continue their climbs into loops as they reversed course to head back out to sea and safety.

They each dived back to very low altitude, and, showing off to their Navy shipmates there on our cruiser, they rolled inverted once more as they flew back over the top of us again.

MiG-21

Even in SupRad, we could very clearly hear the planes as they came in and went out less than a minute later right over us. It was quite a "sound show" and very exhilarating to all of us there in our windowless spaces high up in the superstructure of that high-rise cruiser. (It was called that because it was one of the three tallest cruisers ever built anywhere. I forget how many levels (stories to landlubbers) it was, but I think it was 11 and our spaces were on level eight or nine.

But we all knew we had a job to do, to monitor the North Vietnam reaction. Sure enough, as the sound of the A-6s and A-7 died away, two of my guys, CTI3 (later LCDR) Bob Morrison and CTISN Lenny Moreau, picked up a pair of MiG-21s taking off from Phuc Yen, 72 miles to the northwest of us.

I quickly relayed that information directly to the Commanding Officer, CAPT (later RADM) Thomas McNamara, in CIC via sound-powered phone, giving him the range, bearing, and course of the two MiGs which were below the ship's radar horizon. CHICAGO was in a hard a-starboard turn at that moment, avoiding the shore fire directed at us.

The Captain instantly recognized that if he continued the starboard turn, his forward missile system, with its two ready Talos missiles, would be blocked by the superstructure for what could be a very crucial minute. He immediately ordered a reverse turn, going from hard-a-starboard to hard-a-port. As the ship swung back to the left and CHICAGO laid uncomfortably hard over to port, we could hear the missile fire control radars, which were just above us on the top of the superstructure, rotating to search for and lock-on the oncoming MiGs. Suddenly there was a thunderous explosion from the front of the ship. We thought the North Vietnamese shore batteries had gotten lucky and had hit us.

Our intercept spaces were at the front of, and well up, the superstructure. An explosion at the front of the ship was not good news at all, but then we heard the outbound doppler of a missile launch, followed very quickly by a second such explosion and then down doppler, indicating we had fired two missiles. We were firing over our left shoulder the instant the ready TALOS missiles came to bear on the MiGs from behind the superstructure. The fire control radars had already locked on the MiGs as the superstructure did not block them, so the missiles launched the second they cleared the blocking superstructure.

Their fiery blast burned the paint off our spaces as they streaked away from CHICAGO for Phuc Yen and the on-coming pair of MiG-21s.

Less than a minute later, the first Talos missile intercepted the lead MiG 48 miles downrange. My team detected the reaction of the pilot of what we think was probably the 2nd MiG. It was just a single scream. The first pilot may never have known what hit him. He definitely did not transmit anything. Telemetry from the first missile showed an explicit kill at almost 50 miles. The second missile arrived on the scene a few seconds later. Its telemetry reported seeing nothing but "A large cloud of aluminum confetti with no significant speed," as one of the missile techs later put it. We believe the two MiGs were climbing in formation and that the first missile, with its near 500 lbs. warhead, HUGE for an anti-aircraft missile, hit the lead MiG. It was carrying a full load of fuel and missiles.

USS CHICAGO (CG-11)

It exploded in a massive fireball and debris field, instantly engulfing its wingman, who was also fully loaded with fuel and weapons. It still rankles a bit that they gave us credit for only one MiG. We know from SIGINT that neither of the MiG pilots was ever heard on the air again, and we heard no rescue operation for a downed pilot got underway on the other side, so we are sure we destroyed both MiGs.

All of us in SupRad were overjoyed. I may be more than anyone else. Well less than one hour on watch and in charge for the first time in my professional career, I have played an active, crucial part in a MiG kill. It was a genuine team effort. It was the first "kill" for all of us. But our celebration was going to have to wait as the day was just beginning. We were still close to the harbor, and anything could happen.

IL-28 BOMBER

We cheered for less than a minute. MiGs, IL-28s bombers, and torpedo boats were still very much a threat, and we went right back to our duties very quickly. I was especially concerned about the eight IL-28 medium bombers North Vietnam was known to have. We were so close to the shore that they could launch from any one of several bases within 100 miles of us and hugging the ground, never getting high enough for our radars to see them until they were very close and about to lay a string of bombs on us, or worse yet, launch a torpedo into us.

But that was not what caught the team's immediate attention. One of the operators picked up a radio transmission in a language he did not understand and asked our chief, CTICS Spry, to try to identify the language. The chief listened for a few seconds and handed me the earphones, saying this appears to be a military transmission in the clear in a language none of the team knew. It might be Russian. He knew I had recently attended the Russian course at the Defense Language Institute. Maybe I could help? I put on the earphones and quickly recognized it as a conversation between two Russian merchant ship captains. They had SECRET contingency guidelines from Moscow and had just been ordered by Moscow to implement specific sections. The captains were discussing what they had been told to do and what their plans were to carry out their orders. This conversation was on the radio channel that is known the maritime world over as Channel 16, Harbor Common! Talk about a classic breach of security!

When the transmission was over, I went back and listened to the beginning of the tape, making notes and translating for the chief as I did so. When I finished re-listening to the entire less than 10-minute conversation, my immediate reaction was, "Wow! The President would like to know right now what I now know!" and told the Chief so. I forget which one of us said "CRITIC" first, but I know that was my immediate thought. The Chief pointed out it was my responsibility and call, but he would back me up, either way, CRITIC or SPOT REPORT, but it sure looked like a CRITIC to him. "Me, too!" was my response.

For the reader who is not part of the signals intelligence (SIGINT) community, a CRITIC is a FLASH OVERRIDE message addressed directly to the President via the White House Situation Room. It is reserved for only the most urgent messages with information for the President's immediate attention. It must be derived from SIGINT, most generally specifically communications intelligence (COMINT).

That morning there off Haiphong, I could clearly remember my instructor back at Fort Meade about six months earlier telling my class that "If you are willing to personally wake the President from his first sleep in 48 hours to tell him what you have just learned, send a CRITIC immediately. Even if you are embarked on a warship, do not wait to get the commanding officer's permission. Send it immediately." This was precisely the situation I was in at that point. Indeed, my instructors had stressed that a CRITIC was the sole responsibility of the SIGINT community authority embarked. That was me and had been for something like all of two hours. The CO was nine decks below me and had no real say in my judgment that I needed to tell the White House something. This was my responsibility, and my responsibility alone as the SIGINT authority on-scene.

"In for a penny, in for a pound. This information clearly meets CRITIC criteria." was my immediate next thought. So off went the CRITIC. Yes, I did sweat blood for the next couple of hours. I did seek out the CO to tell him I had sent a CRITIC based on my personal analysis of some information we had intercepted. He was not pleased I had sent a message to the President without asking his permission, but I had just helped him shoot down his first MiG, so he was in a good mood. However, I did stress that I had used my address as a signals intelligence unit, not as a crew member of the USS CHICAGO, just as I had been taught to do. He still was not pleased.

The cherry on the top of this sundae was the "Personal For' to me from the White House that arrived a couple of hours later. The message was from Henry Kissinger, the National Security Advisor for President Nixon at that time. It read, "LT Thomas, Thank you. We needed that information!" /s/ Henry K." And he had info'd the entire Intelligence Community and Navy chains of command.

My team saw it before I did. None of them had ever seen a "Personal for" message from the White House (Neither had I!), and their congratulations still ring in my ear. I served another 18 years as a NavSecGru officer and needed to send two other very high-priority messages to the White House. Once while flying a reconnaissance mission with VQ-1 over the Sea of Japan when my team detected what might be one of our submarines in tactical trouble, the other was the day Richard Nixon resigned from the presidency. My team and I embarked on a submarine which observed some very unusual behavior by Soviet submarine forces. It appeared that they might be preparing for war. My captain and I thought the White House needed to know about it immediately, so I sent a burst message, but 9 May 1972 was my only honest CRITIC. I have met a few officers that have released CRITICS over the years, but I never met anyone who sent one generated on his personal translation and analysis. Indeed, no one else in the first few hours of having the authority to do so. It was one Hell of a start to my career as a NavSecGru direct support officer.

May 1972 proved to be the most active month of the air war over Vietnam for the entire war. USS CHICAGO had a crucial assignment that month. It was the maintainer of the Positive Identification Advisory Zone (PIRAZ), with the essential task of maintaining the identity and track of all the air traffic over the industrial heart of North Vietnam and the adjoining coast. PIRAZ is tasked with sorting out what aircraft are where, why they were there, what, and how they were doing, and why. This was true for both the good and bad guys' aircraft. Our call-sign was "Red Crown," and you cannot read any in-depth history of the air war over Vietnam without seeing references to "Red Crown."

As the embarked direct support SIGINT team, my guys and I were busy providing the bad guys' location and status to the air intercept controllers, in as near to real-time as possible, for a minute in an air battle can literally be several lifetimes. This was especially critical to the MiGs flying below the radar coverage, as they often were until close to their targets (our airplanes). We had several ways to detect and track them that I am still not allowed to explain here.

During my first few days in CHICAGO, I had spent significant time in CHICAGO's CIC as part of my indoctrination into my new duties. While there, I recalled my previous Vietnam tour in 1968 on USS HORNE. Whenever there was any action, I saw the NavSecGru officer come into CIC and man a sound-powered phone on a long cord. I was not cleared for precisely what he was doing, but I was sure his team was eavesdropping on the North Vietnamese air control system, its pilots, and controllers, just like we, the British, the Germans, and the Japanese all did in World War II. Why that was such a big secret remains a mystery to me to this day.

Chicago's CIC was nine levels (stories) below SupRad, so using a long extension cord to connect the two was not practical. However, the first thing I did as we moved away from Haiphong was to seek out the Commanding Officer. I asked if we could install a dedicated secure phone line between CIC and SupRad. I would move my combat station from SupRad to CIC any time we had significant hostile activity.

I explained that I understood everyone in CIC was very busy, but my team in SupRad needed better insight into the tactical picture. Putting me in CIC would allow me to pass that information to them. I would also have immediate access to the electronic intelligence (ELINT) generated by the WLR-1, an overly sensitive passive receiver in CIC. I thoroughly understood the WLR-1's usefulness from my previous experience off Vietnam on USS HORNE, plus the fact that ELINT was a significant part of my submarine duty training.

Being in CIC would allow me to fuse its ELINT with the COMINT being generated in SupRad, thereby creating signals intelligence (SIGINT). Stationed in CIC, I could also be the two-way relay between CIC and SupRad. Being in CIC would significantly shorten our analysis and reporting timeline, thereby producing a much more tactically helpful input. This was especially important in the air battle over North Vietnam.

Exactly how good we were at bringing in all the information available in the electromagnetic spectrum was, and still is, a closely guarded secret, and well it should be. But an open, clear voice? Give me a break! If we were not doing it, people should be court-martialed for dereliction of duty. It is the obvious thing to do. Time is THE critical element in combat. This is especially true in jet air combat, where timelines are significantly compressed due to the speeds involved. I needed to be where I could get all the pertinent information, fuse it, and provide it to the one person who needed it most, as soon as possible. This was generally the Air Intercept Controller, but sometimes the Commanding Officer. That new circuit would allow me to do exactly that.

Additionally, having spent five months in CIC on HORNE off Vietnam, I was very familiar with the radars, communications, and the active electronic warfare (EW) systems, our active jammer suite, and the WLR-1. On USS CHICAGO, there was no one man dedicated to correlating everything, including all radars, Navy Tactical Data System (NTDS) reporting, and EW (COMINT and ELINT).

The Commanding Officer (CO), the Operations Officer, and the weapons coordinator all had their specific jobs. They all had many other things on their mind other than the passive electromagnetic spectrum and what it was telling us via SIGINT. I believed the EW side of things (ELINT) was being short-changed from my CIC time on HORNE, and I saw the same problem here. The ELINT operator or his supervisor had to get someone's attention to report anything, and then it was verbal.

There was a lot of peripheral information to be gleaned from it that was not being fully fused into the calculation as to what is happening.

To make matters worse, in SupRad, we did not know when the coordination team was busy (as was usual during hostile activity) or could be interrupted to report information of widely varying timeliness requirements. The Air Intercept Controllers (AICs) were the guys who really needed the information as soon as possible. By going through the coordinator or the CO, or any other officer, a time-lapse and filters were introduced.

When I asked the CO if I could take on that role during all times we had planes over the beach, he was delighted. I would be on a dedicated circuit to SupRad and pass information to and from it to ensure all in both places, CIC and SupRad, had all the pertinent information available. 38 years later, he still recalled how useful it was to have his SIGINT guy there in CIC with him, and I know the guys in SupRad really liked being updated on the tactical picture in near real-time.

It took the Interior Communications Technician less than an hour to set it up the day I requested it, and it was not a day too soon. Indeed, this proved to be provident timing. Events on 10 May clearly showed my being in CIC significantly smoothed the timeline from signals intelligence collection, COMINT and ELINT, to its processing, to the SIGINT being presented to the man who needed it most at that moment (sometimes it was the AIC, sometimes the CO, in a manner that was easy to understand and digest. So much activity happened on 10 May that, by mid-afternoon, all I needed to do was lean over an AIC's shoulder and put my finger over a place on his scope and hold up 1, 2, or 4 fingers, indicating how many MiGs we thought were there.

By dragging my finger in the direction of travel, he also got a good idea of where to look for hostiles appearing on his scope without me saying a thing. He might look up and ask: "Type?" If I knew it, I would report in a single word. "17", "19" (rare), or "21". We also used a color code of Red, Blue, and Yellow. My being in CIC also greatly assisted in the fusion of all sensors, particularly those gleaned from our electronic warfare suite and the several radars.

The following 36 hours after the morning of 8 May were much calmer. I remember we had subsequent MiG activity on 8 and 9 May, but I remember they were not busy. Nothing like the excitement of 8 May. We might have directed an Air Force and/or Navy plane into intercept(s) on the 9th, but 10 May was so memorable that the afternoon of 8 May and all of 9 May just fades. 10 May was the most significant air battle of the entire war.

It started with the Navy first. They sent in four F-4s as a

MiG Combat Air Patrol (MiG CAP) in a preliminary sweep into the north. Right behind them was a three-carrier "Alpha Strike," over 100 fighter/bombers (F-4s, A-6s, and A-7s).

Often in jungle areas, there are humid nights, hot days, and lots of dew. As the sun rises, its rays heat the very damp jungle canopy causing steam to rise. As the steam rises to a certain altitude, it cools and condenses back into moisture, forming a dense, electronically impenetrable ceiling above the jungle. This phenomenon acts as a waveguide that takes an electronic signal generated on or near the ground and bounces it along its path for many miles in all directions.

So it was both 8 May, which is why we had such early warning on the MiGs launched against the miners, and again on 10 May. As the first four Navy fighters approached North Vietnam's shores, we could hear two MiGs taxing and preparing to take off at Kep airbase near the shoreline. I passed the information to the AIC that "there are two MiGs active on the deck at Kep". I had meant "On deck" in the baseball sense of getting ready for action, but he misunderstood me and relayed the info to the F-4s that there were two MiGs airborne at low altitude near Kep and ordered them to intercept them even though the MiGs were below our radar horizon.

The F-4s arrived overhead Kep looking for the two MiGs they thought were airborne in time to see my two MiGs taking off and dived to engage. They were able to shoot one of them down immediately. They chased the other MiG literally through the treetops until the two Phantoms ran out of missiles to shoot at it, and some of his fellow MiG pilots came to his aid and attacked the F-4s, which went to afterburner and streaked for the Gulf of Tonkin and safety. That 2nd MiG pilot was no fool, and he must have known that if he stayed right at treetop level and kept turning, no one was going to get a clean shot at him.

Later in the day, all I had to do was lean over the correct air intercept controller's radar screen and point to where we thought there were MiGs airborne. I passed similar information many more times that day and over the next six weeks. My team also generally knew their initial course, if not their specific direction. Still, even today, many years later, I am reluctant to say how we did it. Some of this information came from the MiGs' vectors to fly given by their controllers, others from other sources, such as the bearing shift from the ELINT intercepts of the MiGs' radars. Other sources are more classified, and all I can say is that we did it, and it proved particularly useful that day and for the rest of my 6+ weeks on CHICAGO.

By the time 10 May was over, we had counted 41 individual MiG pilots active. Some of the MiG pilots had flown their mission, returned to base, refueled, and went up again.

My team and I had provided early warning to both Navy and Air Force F-4s directing them into or warning them of multiple engagements, including the lead elements of both the USN and USAF initial strikes.

The US had over 335 aircraft airborne that day. At the height of the air battles that morning and early afternoon, the radar screens looked like someone had spilled a shaker of salt over them. There were so many white dots, the radar returns, the various aircraft's reflections, you could not count them. Our team kept inputting needed information to the AICs, but there was so much of it at the height of the battles that day that it was hard to stay totally current (or even coherent!)

The rest of the month and early June were also extremely active, if not as hectic as 10 May. There are four other engagements that month that I still clearly remember. One was when we figured out exactly which two Navy F-4s had a MiG-21 climbing from directly below them, intending to shoot them down. We got a specific warning to the Navy pilots in time for them to roll out of danger. Once warned, the Navy pilots were able to turn the tables on the MiG. Because the F-4s had a significant height and speed advantage at that moment, they were able to "roll over" on the MiG to see it "Standing on its tailpipe, almost out of energy" (meaning it was not flying very fast) "and shoot it down within a very few seconds." as the engagement report read. From our warning to the F-4's rapid role reversal from hunted to hunter to shooter was well less than a minute, maybe even less than 30 seconds. It was very cool to see teamwork in action.

Another time we detected two MiG-17s preparing for takeoff from a specific field where we had F-4s nearby. I believe they were Air Force this time. The F-4s shot both down, but they claimed they never saw the MiGs we were vectoring them against because we could not possibly have seen the two MiGs they had shot down. After all, they were just a few feet off the ground! Our radar could not possibly have seen them…and it had not. My team and I had, and we had, by that time, established such a close working relationship with the AICs that they were willing to believe what we told them such that we could pull something like this off.

MiG-17

We could not explain all this to the AF jet jocks, so we just smiled and went on our way. It still annoys us that our AIC and the CHICAGO did not get credit for those shoot downs. We also ran into the Navy crew at the Cube Club in Subic Bay that shot down the very first MiG on 10 May. They, too, told me that they had missed the first two MiGs we sent them after, and "we could not have known the next two were taking off as they were well below radar detection range." This was true. They were not detected by radar! But in those days, I could not explain how we had done it, so I just kept my mouth shut. At least in the Navy case, CHICAGO got credit for assisting in the MiG kill.

On May 18, we detected what we took to be a pair of MiG-21s taking off from Kep. We vectored a pair of F-4 off the USS Midway to intercept them, which the Phantom did and, after a significant dogfight, ended up shooting both down. I subsequently saw a news report congratulating the F-4 crew for their kills. That is how I learned that the radar intercept operator (RIO), who had visually acquired the MiG, allowing his pilot to position their plane for the first kill, was LT Oran Brown, my first roommate at OCS.

MiG-19

Oran was also a previous enlisted man. They separated us as roommates after the first room and uniform inspections. We both "knew the drill" and aced those inspections—the only ones in the company to do so. Our Company officer decided to "share the wealth," and we were each assigned to room with guys that had failed those inspections. Still, we both graduated near the top of our class. Experience does pay.

Small world, but the story has yet another twist. At our 50th class reunion, I learned that we had misidentified the aircraft. We had said they were MiG-21s because that was the only plane that used an afterburner and was usually operated from that base. However, the F-4 crew saw that they were MiG-19s, which were not known to have ever flown from that base. The MiG-19 was an excellent fighter, but it was difficult to maintain. The only facility equipped to support it was Yen Bai, well off to the northwest.

That brings me to the high point of mid-May. RDC Larry Nowell, the lead Air Intercept Controller on USS Chicago, became the first AIC to register five successful intercepts, thus becoming an ACE. He was flown to an aircraft carrier and told he was nominated for the Distinguished Service Medal and promoted to senior chief. When he returned to CHICAGO that evening, he came straight to SupRad with the news and said, "I owe you guys at least half of this medal. I could not have done it without you."

My team and I felt immensely proud. (& we all agreed with Chief Nowell — it took a team, and we were very clearly part of it.) However, Chief Nowell was an acknowledged master of his craft and was the quarterback, "calling the plays." He became a legend in the Navy. We have stayed in the shadows (until now!)

Another engagement I also distinctly remember as having a few odd twists. My team and I had discussed what we would do if we ever detected a MiG low on fuel as "just in case" contingency planning. I had brought it up at a training session because I remembered hearing discussions among naval aviators about pilots having to bail out due to running out of fuel in a dogfight. Sure enough, about a week later, one of my linguists reported that a MiG-17 had stayed engaged in a dogfight too long. The MiG had reported being in a low fuel state and requested to make an emergency landing at an airfield other than his own. My guy was able to figure out where he was headed. In checking with the AICs, I learned that we had ready F-4s within range of that airfield and suggested we try to cut the MiG off. The AIC was willing, even though he had no radar contact on the MiG.

Sure enough, the airfield detected the oncoming Phantoms and diverted the MiG-17 to another field. We quickly asked the AIC to shift his F-4s to the other airbase. At that point, the North Vietnamese controllers, seeing our Phantoms change course, vectored the MiG to yet another airfield. As the MiG approached the third airfield, he reported his engine had quit as it was out of fuel, and he was going down. The controller replied. "Understand. Keep calm. Check well. Bailout." We had scored a kill without firing a shot! We were jubilant, and I reported the kill to the Captain. He was delighted, too, and congratulated us.

There was a downside to this. The USAF reported the same callsign active later that same day, and our kill was disallowed. I still maintain the pilot had bailed out near a base, the North Vietnamese were noticeably short of experienced pilots by then, and he was asked to (or told to) go back up as soon as he could. We did destroy the MiG but did not kill the pilot, which was OK by me.

Unfortunately, the Captain was less than pleased. "You intel types are always second-guessing each other. That is why I do not trust the lot of you." Those words sting to this day. One day you are a hero, the next, a goat. Years later, he told me he does not remember saying that and that he has a great appreciation for the contributions of Intelligence to the operational forces, which makes me glad. However, my memory is very clear, as those words were engraved in my heart that day, and they still sting to this day.

To give the Captain his just due, we were all very near exhaustion by that point, what with the unparalleled pace of air combat ongoing in his area of responsibility. I am sure he, with the stress of command, was under more pressure than most of us.

The other MiG engagement I very clearly remember did not have a happy ending. We detected a MiG-17 being vectored against a specific airborne US target, but there were many USN and USAF aircraft over North Vietnam at that moment. We issued a "MiGs airborne" warning, but we could not immediately determine where the MiG was and who his target was. We finally figured out that it was a single Navy A-7 on an "armed reconnaissance" mission over the Ho Chi Min trail and passed the info to the AICs, who issued an alert specifically to that A-7. His reply haunts me to this day.

"Thanks. I was wondering where all this horizontal 20 Mike-Mike (cannon fire!) was coming from!" Classic naval aviation cool! But it was his last transmission. We searched for him for several days but never learned where exactly he had crashed, and he is now listed as "Killed in Action." I needed to have been 30 seconds faster to have most probably saved his life. That lesson has stayed with me for the rest of my life.

By the time it was all over, and I left CHICAGO to return to Japan, I believe I had participated in 28 MiG engagements, at least 16 of which were successful. CDR Dreshler, the man who had asked me to join his direct support team at NavSecGru Activity, Philippines, told me that I was the first NavSecGru officer to participate in more than four shootdowns, thus making me a SIGINT ACE. The war went on for another six months, and it is entirely possible another NavSecGru officer became a SIGINT ACE, but I never heard anyone claim to have done so. Indeed, I rarely spoke of it as the guys in the F-4s were the real ACEs.

I remember being off Haiphong and Vinh watching in awe as the night raids of F-4s, A-6s, A-7s, and F-105s went over the beach. A series of glowing orange fountains erupted as the antiaircraft guns and missiles rose from the beach. You had to have "brass ones" to fly into North Vietnam and that orange wall of fire around Haiphong and up the Red River to Hanoi. My hat is off to those USN, USMC, and USAF aviators. I was proud to have served with them. I am also proud of what my team accomplished there on CHICAGO.

I also rarely spoke of the fact that I was apparently the first "ACE" of the Naval Security Group because, at least in part, there were many green eyes when I got back onshore, especially at Misawa. Some of my fellow officers were clearly jealous. I had fallen into a pit of slop and came out smelling like roses. That slop pit had several pools:

1. I was trained to be an NSG direct support officer in submarines and was redirected to aviation. (*In naval intelligence service in submarine surveillance is #1 by a mile.*)

2. I was redirected to the USS WORDEN, a surface ship going into the Sea of Japan. (*In naval intelligence, aviation is #2, surface ships are a distant #3. Demoted twice!*)

3. WORDEN was redirected to the Gulf of Tonkin, and I was along for the ride at that point. (*Shanghaied! Not even going where I was trained to go. What a bummer!*)

4. I chose to volunteer to be a CIC watch officer and train to be an Officer of the Deck. (*Why work a shift job when you do not even really have to work? Are you stupid or crazy? Or both?*)

5. I chose to volunteer to stay in Southeast Asia after spending almost two years training to conduct surveillance of the Soviet Union.

("*Let me get this straight! You volunteered to stay away from home for three more months? And you also volunteered to go back into a war zone where you had just had a man die in your arms from a grievous wound, with many others seriously wounded all around you? You really are stupid! (And maybe crazy, too!*")

What else could go wrong? My ship gets hit with friendly fire, and I get wounded? *Yea, that too!*

I made a series of decisions that were, in hindsight, the right ones, but it certainly was not clear at the time. I can only guess Someone really was looking out for me.

"Lucky bastard! He does not deserve all this fame and good (sic) fortune!" seemed to be the attitude of some of the NavSecGru DirSup officers I had left in Japan.

The fact that I had spent several years training in high-tech naval combat never seemed to cross almost anyone's mind, but then at Misawa, we had over two dozen guys who had come through the same pipeline as I had, and we were all competing for promotion. Some of us, myself included, were reserve officers and were also competing to be accepted into the regular Navy during a time of cutbacks. LT Norm Jewell, my sponsor, did understand, and I much appreciated his friendship and support. Do so to this day!

In looking back, my one real advantage over these other guys was that I had spent five months as an interested and eager young ENLISTED guy in CIC on USS HORNE, a missile cruiser off Vietnam. It was OK for a 3rd class petty officer to ask many questions and make a friendly pest of himself. Had I been a junior officer, I am sure I would have been much more circumspect as officers are just supposed to know a lot of this "stuff" already. This was true even when it is all brand new to everyone except the data systems technicians and radarmen (now called operations specialists).

HORNE also had the added advantage by having the newest and best Navy Tactical Data System (NTDS) afloat. My best friends were two groups that were stationed in CIC with me, the radarmen operating the WLR-1 electronic warfare system and the data systems technicians running the NTDS. Both of their workspaces were close to mine, and both were enormously proud of their "toys." They happily answered all my many, many questions. I even enrolled in Computer Science 101 and Fortran, being taught by CDR Kleber Masterson, HORNE's XO, the second in command. He was selected for his job because he was one of the first graduates of the Naval Postgraduate School in Computer Systems and was eager to teach this new field. I also became good friends with the NSG "spooks" who came and went, and that is the rest of this story.

I now realize I was well prepared to jump into that "pit" of very odd circumstances and do a good job. At the time, I just shrugged off the snide comments as best I could. Jealousy is so diminishing. Let us get back to work!

Alone, Unarmed, and Unafraid
(Hey, Two out of three is not bad!)

Chapter Six

I arrived back at NavSecGru Activity Misawa in late June 1972.

Shockingly, I was immediately assigned Command Duty Officer (CDO) for the first weekend I had had with my family in over four months. Welcome home! I asked the Senior Watch Officer (SWO), the man who controlled the watch assignments, why I was not given a bit of a pass, having just returned home from three months of combat duty.

His answer was right out of MASH. I had not served a weekend watch since I arrived onboard the command five months earlier. The fact I had been shuffled off to a school I had not asked for within three weeks of arrival and then deployed to a combat zone and had spent something like three months at sea in those four months meant nothing to him. I saluted and did my duty which had to be on base, away from my family, which lived off base. Because we had no phone at that time, I had to take a room at the BOQ. It was genuinely rural Japan in those days. My wife was furious, to put it mildly. I was none too happy either, but you do not get ahead by making waves.

However, as wives do everywhere, my wife talked to the other junior officer wives, and word got back to the Operations Officer's wife. I believe via Hugh Doherty's wife, who had, because Hugh was the team lead on WORDEN, taken my wife under her wing.

The Operations Officer was the boss of the SWO, and she told her husband our story. The Operations Officer immediately set a policy exempting DirSup deployers such as myself from CDO for at least a week for every month deployed after returning from a deployment. Indeed, it may have been two weeks. At nearly four months, mine was one of the longest ever for the command. However, it was nothing compared to what the aircraft carriers, cruisers, destroyers, oilers, stores, and ammo ships off Vietnam were enduring, so I had not complained, especially to anyone up the chain of command. But the SWO thought I had and was very angry. I had made an enemy. Not the last time I would make a real enemy through no fault of my own. I will get back to that in the chapter on the creation of Satellite AIS.

At Misawa, I settled into the training schedule for flight duty. I was also training to be an Operations Watch Officer (OWO), the man who ran the command's day-to-day mission, overseeing the collection, processing, and report writing, plus the communications functions. It was a fascinating mission, the details of which remain classified to this day, and well they should. In those days, we could not even allude to the fact that we collected signals of any sort even though the site was festooned with antennas of many sizes and types. The antenna field's queen was the AN/FLR-9, a large ring of antennas many yards across (100+) consisting of at least three circles of different antenna sizes. Anyone could see it had a direction-finding mission in multiple frequencies, but how good it was is probably still classified.

We called it the "Elephant Cage" and often joked we would play football inside of it. It was that big. The operations building was also huge. Not only did it office the hundreds of Navy men of NSGA Misawa, but the Army and the Air Force also had 100s of men working in it as well. It was the most significant US signals intelligence (SIGINT) site in the world. I believe it was the largest in the free world. The Russian SIGINT site at Lourdes Torrens, Cuba, targeted against the USA, may have been the only one bigger anywhere. Most of the men and women in the Operations building were in one of the four US armed forces.

They were involved in collecting, processing, and reporting SIGINT.

NSGA Misawa's 5th division, my division, was a bit different. Its mission was to provide DirSup Teams to ride the ships, submarines, and reconnaissance planes to deliver intelligence directly to the warships and aircraft of the US Navy's 7th Fleet. It was responsible for manning all units deployed to the western Pacific north of Taiwan. I say "intelligence" rather than "signals intelligence" advisedly; we were THE technical intelligence collection assets at sea.

Besides becoming qualified as an Operations Watch Officer (OWO), my primary task was to get ready to deploy to NavSecGru Detachment Atsugi (NSGD Atsugi) for duty onboard Fleet Air Reconnaissance Squadron One (VQ-1) EC-121M and EP-3B aircraft. VQ-1 also flew the EA-3B, but they were assigned primarily to forward-deployed aircraft carriers, and in any case, I was not yet qualified to fly in them. Only our best, most experienced men flew in them, and I was just a rookie at that point. A rookie with some extraordinary experience, but still just a rookie as far as the senior officers at NSGA Misawa were concerned.

Misawa Air Base Operations Area
Antenna, Operations Building, Barracks, Gym, and Chow Hall

About seven weeks after returning from the Gulf of Tonkin, I was judged ready, after several examinations, both in writing and verbally, to go to Atsugi for flight duty. I was hyped! If I could not serve in submarines on secret spy missions, I wanted to fly on secret spy missions.

I drove 800 kilometers from Misawa, located in northern Japan, to Atsugi, about 45 kilometers south of Tokyo. In those days, no freeways connected northern Japan to Tokyo. However, off to the west of my route south, I could see north Japan's first freeway under construction. But I was on local surface roads with plenty of stop lights and signs for nearly the whole way.

While the trip took two days, it was an adventure in and of itself, and I loved it. If you are polite and know even a few words in their language, the Japanese will fall all over themselves to be friendly and help you any way they can. My many trips into the real Japanese world off base are still some of my fondest memories, and I count many Japanese as real friends.

Once at Atsugi, I met the team, both the NSG guys and the VQ-1 squadron guys, flight crew as well as support staff, including the intelligence and maintenance officers and the two civilian technical representatives, the legendary Chuck Christman and Elmer Achtenberg, both sadly, no longer with us. May God rest their souls. They both deserve unique places in Heaven. Both had been in Japan a long time, since the early 1950s. Elmer was an electrical engineer and mathematical genius. Chuck was a true "out of the box" thinker, an ideas guy with a vast amount of technical knowledge.

Many of the unique pieces of intercept and analysis equipment in the VQ aircraft were either invented or made better by these two men, especially Chuck. When I met him, 1972, he had three Navy Civilian Distinguished Service Medals. I think he received at least one, probably two, or maybe even 3 or 4 more. At 80+, he was still working at the Naval Research Laboratory doing "leading edge" work. He was a super mentor and remained an idol of mine and many others.

I would have a great deal of interaction with these men, especially Chuck, during the rest of my six + years in Japan and, in the case of the VQ-1 officers, their reliefs.

The officer who stood out in my memory is Joe Mazzafro. He was the last intelligence officer I served with there. He had more energy than nearly anyone else I ever met. Indeed, he is widely known and respected at all intelligence community levels, including the very top, to this day.

Nearly 20 years later, he and I worked together at Johns Hopkins' Applied Physics Lab. I believe the two of us had more operational intelligence experience than any other four, maybe even six, people at APL at that time. There were many retired members of all four services working there. Still, we were the only operational intelligence types at APL until toward the end of my time. Joe hired a marine intelligence officer as his assistant, and another NavSecGru DirSup officer joined our division. So, then there were four of us, out of over 3,000 people employed there.

EC-121 OVER Mt Fuji

Two or three days after my arrival at Atsugi, I went up on

my first mission to the Sea of Japan. VQ-1 flew reconnaissance missions routinely four or five days a week. If a significant event was on, such as a Russian operational test of a weapons system or an exercise, we flew every day during the event. Sometimes VQ-1 even puts two birds up to cover exceptionally high-priority actions. My first flight happened to be during one of those high-priority occasions. We successfully covered the test of a new weapons system.

I was hooked! These flights were exciting and of enormous value to the security of the free world. Generally, people say things like this as a joke, but this was no joke. We gained vast amounts of valuable intelligence. The intelligence we were collecting would be invaluable if we ever did get in a shooting war with Russia or any of its many surrogates. My allegiance to the submarine corps began to waver. I also loved my flight time with the USAF. Very few airmen knew anything about the value of submarine surveillance, so the aviators ruled supreme in that service. I enjoyed that very much, but I am getting ahead of myself.

I had many extraordinarily successful missions in the Sea of Japan, over the DMZ in Korea, and the Pacific off Petropavlovsk, but most of the details of those missions are still classified. I recall odd incidents, like when the pilot of an EC-121 called me up to the cockpit to identify an odd-looking aircraft. By the time I got to the cockpit, the plane had vanished, so I said something to the effect that it "must have been a phantom," as a bit of a play on words, the F-4 Phantom being our first line fighter at the time.

The mission commander overheard my remark and said: "No F-4 has the range to be this close to Russia; it must have been a Soviet fighter. We must issue an alert that we have been intercepted." This is how misunderstandings take place! The pilot and I had to explain to him we had not been intercepted by a fighter but instead had seen an unusual-looking aircraft. It was probably a transport that had just passed by us, and I was trying to make a play on words as a joke.

On one mission in much the same area, we did have what appeared to be an impending intercept by North Korean MiGs while well out to sea south of Vladivostok. We dove for the deck and pulled out so low a wave hit the radar dome on the bottom of the EC-121. At that point, I think I was more scared of us crashing because the pilot was going to make a mistake and fly us into the ocean.

We also had a near mid-air collision with a Boeing 747 SP (Special Performance), a shortened version of the regular 747. Because it was shorter and thus lighter, but with the same engines, at that time, it could climb and fly faster and higher than any other airliner in the world except the Supersonic Transport. We were on a mission over northern South Korea. Our track took us right over Kimpo International Airport in Seoul. We thought we were high enough that no airliner could reach our altitude, but no one had considered the climb rate of a lightly loaded 747-SP with an enthusiastic pilot.

Workstation/Position in EC-121

At the time, all I knew was that I was standing in the mission area of the plane flying at 20,000 feet, when we went into a steep dive. I caught myself just in time to prevent a very nasty fall from the rear to the front of the mission area, a length of something like 50 feet. It would have been a very ugly landing from that fall. Our pilot came on the intercom immediately after we regained level flight and apologized for the dramatic maneuver, but we had just missed a head-on crash with a 747 SP.

Thirteen years later, I was on an American Airlines flight on the great circle route from Dallas to Tokyo. That route takes you right by Russia's east coast. As we flew by the Kamchatka Peninsula, well out to sea in international airspace over the Pacific Ocean, I was looking out the window. I remarked to a flight attendant that I had seen these mountains sticking up from the fog many times, but this day was exceptionally clear, with glistening clouds. They looked like the pure white sand at Ryoanji, the famous Japanese Zen garden at Kyoto. The mountains poking up through the clouds looked like the stones at that elegant and peaceful garden.

I mentioned to the flight attendant that it was the most beautiful view of the peninsula I had ever seen. She asked if I often flew this way and explained I had flown in this area in USN and USAF reconnaissance aircraft. She told me the pilot in command of our flight had told them the previous night at dinner about his having a near mid-air collision over Korea about a dozen years ago with a Navy reconnaissance aircraft. Maybe I knew something about that? Wow! Talk about a small world!

She went to the cockpit to tell the captain about me, and a few minutes later, he came down the aisle and sat down in the empty seat across the aisle. After briefly introducing ourselves, we exchange our sides of that story. He had a very lightly loaded 747-SP that day and decided to see how quickly he could get to altitude. He had asked for and obtained clearance to go straight to 40,000+ feet, but no one had considered he would climb fast enough to almost hit us. We swapped stories for some time. It was a very memorable flight for that reason, as well as the beautiful scenery of the mountains of Kamchatka poking up through the clouds glistening in the early morning sun.

My most memorable flight in an EP-3B was also along the Korean DMZ in 1975. We were headed east, and there was such a strong headwind that the pilot decided to see if we could actually fly backward! He kept bleeding off throttle until we were just creeping along at about 10-15 miles an hour, the powerful headwind holding us aloft. We finally got to the east coast of Korea, and we needed to turn around and head west, back across the Korean peninsula. As the pilot initiated the turn, there was a loud noise, not unlike an exploding SAM warhead, to the left of the airplane, followed a second later with an equally emphatic "explosion" to the right, and the EP-3B went into an uncontrolled spin.

My immediate thought was the North Koreans had decided to take a shot at us with one of their surface-to-air missiles fired without any guidance systems on (which we would have detected), and we were the target of their test! Not all that encouraging! My next thought was that this might be the last few seconds of my life, and I thanked God for letting me have such an interesting and exciting life. As the altimeter above my head rapidly unwound from 22,000 plus feet down through 18,000 feet, I could feel the airplane come out of the spin. We were still headed down at a steep rate, but the spinning had stopped, which significantly improved our situation.

The pilot hauled back on the stick, and we came out of the spin and dive just below 16K feet. As we climbed back to our mission altitude, the pilot came on the announcement system and explained what had happened. He had not brought the engines back up to speed before he started the turn and, as we turned across the powerful headwinds, we lost the lift they were providing; indeed, the winds went from helpful to dangerous in the blink of an eye, and he lost control of the plane. The loud noise we had all heard was the wings unloading their lift as the wind shifted across them. He took full responsibility and asked for our forgiveness. Those of us in the mission compartment all agreed we wanted to string him up at that moment. Still, the pilot was one of the best in the squadron, and there nothing was to be gained, so the mission crew guys in the back, both the NSG and VQ-1, just kept our mouths shut. I understand the incident did become a teaching point in the P-3 training syllabus which he went to head when he left VQ-1.

EP-3B

On another mission over the Sea of Japan in an EC-121, I was sitting in the galley preparing all the initial input into my mission report. It was pen and paper in those days, and even simple tasks were time-consuming. The flight engineer and an ensign on his first flight were there with me. Suddenly we all smelled smoke and agreed it was, because of its smell, probably from an electronic source. The flight engineer started checking things out in the galley. I went forward to the cockpit, which was about 20 feet forward of the galley, to let the pilot know there was a potential fire in the galley area.

While I was on the flight deck, the flight engineer located the problem to be the drive motor of the 14-foot-wide antenna of our APS-20 radar, which was in the big belly bulge (15 feet across) below the galley. He pulled the floor planks in the galley, opened the inspection hatch, and crawled down into the radome to fight the fire, which he successfully did. He also stationed the ensign over the hatch to warm me and anyone else coming into the galley of the open hole in the floor.

Simultaneously, the pilot decided, "Better safe than sorry," and ordered the crew to go to ditching stations. He also started a steep dive to get down near the sea to be ready in case he had to crash land the plane in the sea to save us from burning to death. My ditching station was all the way aft, and we were now in about a 40-degree dive. This angle meant I was trying to go up a steep hill, and I was using handholds to drag myself up the aisle as I came to the open-access hatch. The ensign saw me coming and, without saying a word of warning to me, stepped out of my way, and I stepped into the open inspection hatch in the floor. My right boot caught on the access hole threshold, but the only reason I did not break my left leg was due to my left boot hitting the flight engineer with a glancing blow to his head, absorbing much of the impact; however, my left knee has never been the same since then.

Even now, more than 40 plus years later, my knee occasionally locks up; or becomes painful in cold weather. Still, until recently, I could generally walk it off, but now the pain is there all the time. I will be getting a knee replacement.

I also almost castrated myself on the combing of that access hatch that day. Ensigns! He was a particularly hard case; he was smart as a whip technically but must have played hooky the day they passed out the common sense. He left the Navy as soon as possible and became a highly successful engineer in the aerospace world. I saw him several times at trade shows and once when I visited an electronics firm for a live demonstration. He always apologized. I just laughed it off, but I now use a cane.

Another definite "Aw S...!" was when I announced in a post-mission report that the first India class submarine's long-awaited sea trials had commenced. It was a unique submarine with two recesses on its after-deck for miniature submarines. Naval Intelligence was very curious as to how it was going to be used. I was delighted with my team and myself for collecting the very first SIGINT on this unique submarine.

During the flight, I discovered I was the only person onboard the flight who even knew of its existence, which surprised me a bit. It should have alerted me to keep my mouth shut and my pen still. In those days, I was very proud that I read every bit of intelligence I could find, especially on the Soviet Far East and the North Koreans.

When I was not flying with VQ-1, I was in the Operations Watch Officer rotation at Misawa. I had access and encouragement to study material from very high classification levels well beyond Top Secret. VQ and surface operations were generally SCI, just one level above. The satellite collection was a level above that. Special submarine and SEAL operations were two different but co-equal levels above that. I am sure there were and are other compartments and levels beyond that.

Less than an hour after the Post Mission Report (PMR) went out, I received a call from our communications center. The communications watch told me there was a call for me on the secure line from the Pacific Fleet Headquarters watch officer. The center was several doors away from my office, and I hurried there to take the call, expecting it to be one of congratulations.

However, my blood froze when his first words were: "Where did you learn the Russians have a submarine with those capabilities?" I instantly realized I had just stepped across not one but two security levels and could be eligible for court-martial. I think I swore and said, "I just goofed up big time, didn't I!" His reply was, "Yeah, you did. But we are almost finished preparing a report on this new capability and had planned to push it all the way DOWN to Top Secret next week, so my boss said to let it slide. But BE MORE CAREFUL, OK? You can get a court-martial for this sort of thing." It was at that point that I think I started breathing again. I replied, "Roger, Thank You! It will not happen again! Thank you, sir!" and we hung up. Close call!

My name is not Hilary Clinton. I could have been in BIG trouble for accidentally releasing even a single message of that classification on a circuit only cleared to Top Secret/SCI. I can just imagine my punishment for allowing 33,000 highly classified messages to be compromised by putting them on an unclassified, open server; and then blatantly lying about it. I might still be in jail today.

I also flew coordinated missions with the USAF, they in their airplanes, and me in mine. I was fascinated by the USAF. They seemed to have more money and people than God. I studied what they did and how they did it to see if the Navy could learn any new tricks. Most flyers of all services and nations spend a great deal of their time beating their chest and chanting, "We are the greatest." That just seems to go with the mystic of flying, especially regards flying single-seat fighters.

INDIA CLASS SUBMARINE

However, when it comes to flying large reconnaissance aircraft, especially those of the USN and USAF, that is counterproductive. They are likely the best in the world in this area. They both achieve the same goals but in different ways, and they are stronger yet when they collaborate. Several years later, when I went on exchange to the USAF Security Service, I was to get the chance to improve on that collaboration and make it more productive. And I believe I did. Even more importantly, I know my USAF bosses thought I did, as did at least three admirals. It was a dream assignment!

All my time in collection missions in all three environments (surface, submarine, and air) in both the Navy and Air Force was special to me. Each in a different way. I suspect it is like the way a mother loves her multiple children. Equally, but differently.)

My most memorable actual flight in the Navy was during a joint USN/USAF operation in the Sea of Japan against a high-priority Soviet fleet exercise. My crew was in an EC-121, the lowest in terms of altitude, with an EP-3B about 10K feet above us and an RC-135M, Rivet Card with the officer in tactical command (OTC) embarked, about another 10 K feet above that. A Combat Sent (RC-135S) was also on strip alert at Misawa, ready to launch on the OTC's command.

It was already a busy day as we flew just outside of the Soviet-declared test range collecting intelligence, but it suddenly got even more dynamic when one of the linguists reported that a BE-12, a Soviet antisubmarine warfare (ASW) aircraft, on range security duty, had made a positive visual identification of a foreign submarine well out in international waters.

However, what really got our attention was when the Soviet aircraft requested permission to engage and drop a weapon on it. Just like on the USS CHICAGO nearly three years earlier, I looked at my Leading Petty Officer, and he looked at me. But this time, we did not have to say anything. We both knew what I had to do in this situation, and thus I sent my one and only special-purpose CRITIC, then called a "Frost Fish." Like a "regular" CRITIC, it goes straight to the White House Situation Room at the highest possible priority. Luckily, our follow-ups were less dramatic. The Soviet OTC refused to believe the BE-12 pilot, who pleaded with him for a minute plus and then shut up, his plans of earning "Hero of the Soviet Union" gone like the mist. We were relaying each transmission of both the Soviet ASW aircraft and his officer-in-tactical command to the White House via the Naval Communications Station, Kamiseya, Japan.

It is a short distance from Atsugi, our operating base. Later that night, as our flight crew was relaxing at the Atsugi Officers Club Bar, LT Peter Mast, the Communications Watch Officer at Kamiseya that day, walked into the bar. Peter was a hail-fellow-well-met sort of guy (God rest his soul) and saw us as he walked in, booming out: "I thought I would find you guys here."

I answered him back with, "Well, I bet you had a pretty hectic day, too."

"No!" he replied. "I never did like John Brinker anyway." John was my next-door neighbor at Misawa and a more likable guy as you ever hope to meet. More importantly, he may well have been on the submarine that had been detected. It could also have been Chinese, Japanese, or Korean (North or South). There in the bar at Atsugi, we all agreed we would just have to wait until John got back to port and made his trip report before we found out whether it was him or not. Even today, many years later, that is all I can tell you.

I can tell you NO ONE criticized my "FROST FISH." Even though we practiced sending them reasonably often, my message was the first real one I had ever heard of being transmitted from the Sea of Japan. I was now two up on my mates competing for promotion from the reserve to the regular Navy, but we never mentioned it. I now had a CRITIC and a FROST FISH in my record.

My most memorable flight was one I did not take. In October 1973, the Arabs and the Israelis were at it again. With the massive military aid of the Soviet Union, Egypt had prevailed for the first 40 hours. Then the Jews got their act together and destroyed most of the Egyptian defenses all the way to Cairo. Moscow, seeing the potential loss of its most important client state in the Middle East and a great embarrassment to them due to the quantity and quality of the weapons and training they had just provided, started positioning its parachute forces for a jump into Israel. The US national intelligence collection apparatus detected these preparations. The Soviet troops were to be dropped behind the Israeli Army, which, at that point, was on the Egyptian border.

President Nixon and his National Security Advisor Henry

Kissinger were having none of it. They bluntly told the Russians in no uncertain terms that the US considered an attack on Israel to be an attack on the US. If Israel were attacked, Russia could expect multiple attacks from many different directions almost immediately. These attacks would have come from aircraft carriers as well as other sites and forces.

I, of course, did not know all this at the time, and that morning we launched for a routine flight in the northern Sea of Japan. Early in the afternoon, we got a message on the High Command (HiCom) net, which was reserved for vital messages, saying, "RTB NOW!" Return to base immediately! We knew of no reason for the order, and as we headed home, we speculated what could be the reason. As we neared Tokyo, we could hear many USN strike and fighter aircraft on the radio inbound for Atsugi and requesting permission to land. It appeared at least two aircraft carrier air wings were positioned at Atsugi to enforce our president's words. We shortly learned that the planes on Atsugi were to be the spear's blade, and we were to be extremely near the tip!

Our intelligence officer met us at the plane, which was highly unusual, and started talking a mile a minute. VQ-1 was to be tasked to identify the Soviet air defenses as they became active so the "Iron Hand" aircraft would know which sites to attack first. Iron Hand were the "Wild Weasels" of the USN. A-6 and A-7s armed with Shrike anti-radar missiles (like those that hit the USS WORDEN with such effect).

Thus we, VQ-1, with the Iron Hand guys, were going to be the tip of that spear aimed at Vladivostok! The Iron Hand aircraft would be tasked with destroying the air defense radars in Vladivostok's vicinity. This would allow the Navy and Air Force aircraft's follow-on strikes coming out of Atsugi, Okinawa, and Guam to attack their targets without substantial opposition from the surface to air missiles. He ended with, "Go home and get some sleep. We will call you around 4 AM if we are going." Yeah, right! I wonder how many of us slept that night? I think I got something like two hours, myself. Anyone who was not concerned about their life expectancy that night would have had to be drunk.

The story above leads directly into this chapter's title, which comes from a saying familiar to all four services' reconnaissance flyers. "We go into harm's way 'Alone, Unarmed, & Unafraid!'...Hey, two out of three ain't bad!" I am very sure I would not want to fly a recce mission with someone who was completely unafraid. The ones with no fear are the guys that make the mistakes that get folks killed.

When I was not flying at Atsugi, I returned to Misawa and served as an Operations Watch Officer. I was sitting at a unique vantage point. The secondary downlink site for both USAF and USN flights was Misawa. I saw the US Navy reports very shortly after they left the airplane. They came to us via the Naval Communications Station, Kamiseya. The USAF watch center across the hall was linked to the USAF reconnaissance aircraft via Yokota Air Base. The USAF at Misawa would send out the reports from their flight operations to the world, including the Navy when they thought it might be of interest. I decided to make friends with the "zoomies" across that hall, just to understand how they did their business and maybe give them a few clues of what we thought was important. (Like taking courses in computer science, this turned out to be one of the more innovative, and important to my career, things I ever did.)

A few of the crusty old senior sergeants on the Air Force side started wondering why a naval officer spent as much time as I did on their side of the hall, so I decided to write a formal suggestion to my Commanding Officer. Via my Division Officer and the Operations Officer, I recommended a formal agreement between the USN and USAF SIGINT teams at Misawa to allow and encourage visits between the operations center's personnel. Each organization should be welcomed in the other, workload and pace of operations permitting.

To my delight, the idea in my proposal was discussed by the two commanding officers, USN & USAF, and adopted. I now had official sanctions/blessings from both COs to be as nosy as I liked. Of course, one had to be very mindful of the pace of operations and not bother folks who were busy with actual tasks. But it was at those times you learned the most. I was always careful to stay in an obscure corner of the watch center's command module. This was much like I did in CIC on HORNE when I was a junior enlisted man, and kept my mouth shut until the event had passed. Then I had more questions to ask!

EC-121 Being Attacked by MiGs
In remembrance of the Navy, Marines and Air Force aircrew men who did not come back

In early February 1974, the Fleet Support Division officer at NSGA Misawa called me to ask if I still was a volunteer for submarine duty. "You bet!" was my immediate answer. "Good! We need you to start work-up immediately for a mid-March deployment." I was finally getting my chance to do what I had trained to do. I was elated!

Submarines at last!
All I hoped it would be!

Chapter Seven

In March of 1974, I finally got my chance to fulfill my dream of participating in what many people in the national intelligence community saw as the "crown jewel" of US intelligence, submarine "Special Operations." You will notice that I did not say "US Navy Intelligence"; I said "US Intelligence," as that was reiterated to me many times even before I got into the program. I had gone through training at the Defense Language School at Monterey, California; then an extensive intercept equipment operator course down to the core physics level and the basic submarine training course at Groton, CT; and then a series of classes at the National Security Agency, Fort Meade, Maryland, including cryptanalysis, traffic analysis, non-Morse search and development, radar identification and signature analysis.

I also went through the Direct Support Division Officer training course. What to do, and even more importantly, not do, as a crewmember and division leader on board a submarine. The instruction covered both before getting underway and then underway on what amounts to a full-up war patrol. The only difference was that we hoped not to have to fire our weapons. We also went through the training to be Special Security Officers (SSO) to understand what information needed to be protected at what classification level; and how to do so, including how to handle, pack, wrap, mark, and ship each level of classified information.

My first assignment to a submarine was as the NavSecGru assistant Division Officer on USS Pintado, SSN-672, CDR (later VADM) James Guy Reynolds, Commanding. God rest his soul. The NSG DirSup team chief was LT Dennis Sheppard, one of the smartest and most gracious men I ever met. Unfortunately, he later got crosswise with our commanding officer there at NSGA Misawa over his choice of girlfriend, a beautiful and intelligent Japanese girl who became his wife a year or two later, and his career suffered because of it. Denny would have made a great Admiral, but he retired as a commander. I suspect that the only way he made that was because every other CO he ever had rated him very highly, as did I. Eventually, he was one of the original cadres in the Navy's entrance into what is today known as cyber warfare.

The Navy stood up a team to look at what was initially known as "Information Warfare" in the early 1980s, but it was very black, and I know truly little about it except that many of the brightest of my colleagues were assigned to it.

None of them would talk about it at all, even to those of us that had clearances well beyond Top Secret. In our business, when one of your friends says, "I am sorry, but I cannot discuss this with you." And you both are already cleared several levels above Top Secret; it is considered very rude to press. I guess you call it professional courtesy. So, I did not push, but I did imagine and wonder!

Denny was a great trainer, and I learned a lot from him. He fully respected my experience, especially my familiarity with and interest in hostile radars and data systems. Our contemporaries were linguists and cryptologists, as we were, but most of them seemed to feel that understanding radars and data links was almost beneath them for some reason. Denny and I saw eye-to-eye that if it radiated energy, we wanted to understand it and how our potential foes were using it. And even more importantly, how they intended to use it in a war. I had come to this belief because of my curiosity about the WLR-1, the ESM/ELINT receiver, on USS HORNE in 1968. and subsequent experience with the WLR-1 on USS CHICAGO in 1972 off Vietnam. The WLR-1 data on CHICAGO very clearly told me it was a much-changed, more complex electronic environment in 1972 than what I left in 1968.

Denny did not have that experience base but came to the same conclusion by his mental powers and his readings in the open-source technical press. As I said above, he was brilliant.

The plan was for the USS PINTADO to leave Pearl Harbor for a hostile coast off to the west in late March. We got to Hawaii about a week early to check out all our gear and interact with the wardroom, the officers, and some critical senior enlisted men. A team of some of the best maintenance technicians in the Navy was assigned to Pearl Harbor's fleet readiness organization charged with keeping these submarines in top-notch shape. This was especially important because those submarines are, indeed, on the front line, just like Strategic Air Command bombers. I am sure they still are today.

Submarine near Periscope depth

Our first afternoon on the ship, the CO asked the wardroom to meet him at the club to welcome us. Off to a good start! As Denny, myself, and the PINTADO officers got to know each other over a few beers, we talked about our previous tours and experiences. Captain Reynolds mentioned he was probably the youngest CO of a nuclear submarine ever and stated his age, which seemed insanely young. And then he turned to me, looked at my two rows of ribbons (a lot for a "spook" lieutenant in those days), and asked, "How old are you anyway?" I figured he had just shaved about 4 or 5 years off his age, so I did the same. I was 31 but said I was 27, more as a joke than anything else. CDR Reynolds came back like a shot. "Must have been a tough 27 years!" It is the hardest putdown anyone ever got on me!

He was a great leader. His stories from his career were fascinating. He had served as the recorder for the courts of inquiry of both the USS Scorpion and USS Thresher tragedies. Both submarines were lost with all hands. Once underway, he gave us both cases in detail, including the Good, the Bad, and the Ugly.

Indeed, his stories were lessons in leadership and education for a successful career in the Submarine Service. (He went on to retire as a Vice-Admiral.) He was also about the best poker player I ever met. I was not at all surprised he made it to 3 stars. That last one, the 4th, is more luck and timing at that aptitude. Unfortunately, he died of cancer not too long after retirement. That may even have been the reason he retired. It was a significant loss.

We got underway on the scheduled day after a hectic week. We immediately went very deep, just to check things out in that stressed environment. After everything checked out OK, we returned to a more reasonable depth. It was still much deeper than anything at which a World War II submarine could operate. We headed for our patrol area at a high sustained speed. It seemed very natural, but it still gave me a thrill. Once on station, Denny and I stood alternating watches. He had the primary day watch and me the night, but we overlapped about six hours a day. We needed to be at periscope depth for our primary sensors to work, but there was pressure for the ship to stay deep to optimize its sonar capabilities. This is the classic trade-off of a submarine reconnaissance patrol.

Even if I could remember all the significant intelligence we collected, I could not detail it here. However, I remember one significant event very clearly. We established a trail on a Soviet

Yankee-class ballistic missile submarine. It appeared to be in pre-deployment workup, training for an eastern Pacific missile patrol off the US West Coast. This meant we were going to stay deep for some time, so I went to turn into my bunk and get some sleep. As I opened the covers and started to climb into the bunk, there was a distinct "Whump" and a sharp jolt. We had collided with the Yankee! Knowing the tactical picture from a few minutes before, I guessed we had hit it in the stern, and we had.

I jumped back out of my rack and hurried to our spaces just to be ready. CAPT Reynolds had ordered us to periscope depth to allow our antennas to detect any response to our collision. Our antennas cleared the water at about the same time I arrived at Radio. Almost immediately, we heard the Yankee broadcasting in the clear that an unknown submarine had struck it while submerged. It had been hit from behind in its stern and had no propulsion or steering. The shore station immediately acknowledged the call.

That was all CAPT Reynolds needed to know. We "pulled the plug," going into an emergency dive and, while lowering all masts, we went deep and fast to clear the area as rapidly as possible. A couple of hours later, we slowed and returned to periscope depth (PD) to listen to the tactical situation. Luckily, the Soviets were not looking for us anywhere near where we were. So, we continued to clear the area at PD, monitoring the search for us that grew bigger and bigger and, luckily, further and further away. Once we were well clear, we returned to cruising depth of several 100s of feet and proceeded to Guam. Once there, we entered a covered dry dock to look at our damage, which was significant. So ended my first submarine patrol. Not quite the success of my first surface ship and airplane missions, not by a long shot! But I had still learned a great deal under Denny's tutelage, and I was not dismayed. Disappointed, yes! Dismayed, no!

The ELINT and special signals riders went back to Pearl Harbor. The Sonar expert returned to Washington, DC. My team and I, the linguists and radio intercept operators, the I and R-brachers, caught a plane to Tokyo for further transfer to Misawa. It was very foggy in Tokyo, and the landing in that Boeing 707 was more like a crash. It was the hardest landing I had ever experienced. I felt happy just to walk away from it!

It was so hard we swayed from side to side, as we careened down the runway. The doors on the overhead luggage bins flew open. Luggage fell on the passengers, hurting some of them as the plane lurched down the runway, hitting on the left landing gear, then the right, and then back and forth several times. I thought the plane might have even hit its wingtips on the ground at least twice, but I do not know for sure. That would have been when the bags fell out. The pilot reestablished control, corrected the situation, and we were able to disembark in the usual manner. Luckily, none of us were hurt. I had heard many times that Boeing built sturdy airplanes!

Now I know it!

USS POGY (SSN-647)

We returned to Misawa, and I was recommended to lead my own team. As with the military anywhere, once you make the grade, they run you to death. I was asked if I wanted to immediately go into workup for a mission leaving in June. I, of course, readily agreed, and mid-June saw me back in Pearl Harbor as NSG team chief on USS Pogy (SSN-647). Our departure did not go as smoothly as my first trip. Pogy needed to change one of its main generators.

Having worked in an auto repair shop, I was not dismayed. Changing a generator on a car is no big thing, right? I quickly learned changing the main generator on a nuclear submarine is, indeed, a BIG thing. You need to go into a dry dock and cut a huge hole in the submarine's side through two hulls made of HY-80, the strongest steel in the world. After the new generator was installed, we needed to reweld the two hulls and do so in such a way as to allow the hull to withstand ocean depth pressure to over 2,000 feet. Not easy, but they got it done in 10 days.

In the meantime, we mustered the crew every morning and counted noses. We attended to any business and then went back out on liberty. It was a 10-day paid vacation in Honolulu. Who says the Navy is not a good deal? I got to know my team very well and they, me. They were all good guys, but the standout was CTI3 Jimmy Calvery, a "country boy" from east Texas. Smart as a whip but in far too long to be just an E-4. He and I talked about that, and I found out he simply was not motivated toward advancement. He loved what he did and thus was exceptionally good at it. I lit a fire under him, and he made 2nd class on the next advancement cycle and then 1st class in the minimum time. He went on to be an officer with a sterling reputation. Years later, I visited his family in Texas. His father was the director of a steel mill, and his boyhood home, which Jimmy would have had you believe was a shack, was an elegant home. He just loved what he did and saw no reason to follow in his father's footsteps in the business world. Being Texans, his family was immensely proud of him, and rightly so.

We finally sailed, and not a day too soon. We were all about

out of money, but it had been well spent, helping Hawaii's entertainment and bar industry prosper. We were ready for sea and arrived on station prepared to go to work and work we did! It was a very productive mission, and the captain, CDR Dave Cooper, and I got along very well. He was another of those men I was surprised who did not make flag rank. In his case, very surprised. He had it all in one package. The entire crew and I thought he was a great leader.

Our intelligence take was exceptional by all measures. Still, I remember the little things best, e.g., because of my demonstrated extraordinary night eyesight, I was asked to "volunteer" to be a topside lookout when we surfaced at night off the hostile coast. We needed to reapply the defogging solution to both our periscope lenses.

Submarines have two periscopes, both for redundancy and because they have two different functions, surveillance and attack. The attack 'scope" has just optics and is very thin. It makes a much smaller wake and is harder to see. The surveillance scope has several sensors and therefore is much thicker and leaves a much bigger wake.

I had mentioned my unusual night vision in passing one night in the wardroom. CAPT Cooper, who was also very proud of his night vision, put me to the test. He came away agreeing that my vision, especially at night, was exceptional. So, when he needed a good pair of eyes, he knew where to look. Our main periscope clouded over and needed a new coating of "Never Fog." He picked a moonless night and relatively lousy weather and surfaced. I was selected to be a lookout. Up the ladder inside the sail, we went and took our stations. The northern Pacific is a cold bitch even in the summer. I was awfully glad when we secured and went back inside the submarine, which submerged and returned to the port we had under surveillance. I went to the wardroom for a most welcome big mug of hot chocolate. It would have been better with rum, but this was a US, not British, submarine.

Speaking of rum and chocolate, one of my favorite restaurants in San Francisco was the Sea Witch in Ghirardelli Square, which overlooks Aquatic Park and San Francisco Bay. One of my best friends was one of its bartenders. They served "Swiss Navy Grog," which was nothing but hot chocolate and rum. It was perfect for the chilly, foggy nights in SF, and it is still one of my favorite cold-weather drinks. I also really liked the place because it was on the waterfront, and by then, I was totally hooked on all things maritime. Indeed, I still am!

I can still remember sitting in the mess deck that night off Petropavlovsk, drinking hot chocolate that reminded me of the Swiss Navy Grog at the Sea Witch. For just a second, I thought, "I sure wish I was back in SF at the Sea Witch." Then, just as quickly, I thought, "**No, I don't! I am exactly where I want to be!**

An SSN returning from Patrol

Frequently, I was asked to come to the conn, the control center of a submarine where the periscope is located, to look at a contact. Often the watch had been trying to identify for some time. Usually, I could identify the ship very quickly. I distinctly remember my best two and worst single events. The first was when the Ops officer got me out of bed at about 1 AM to come to look at an odd-looking ship. They had been trying to ID for over an hour and had all the recognition books scattered about in Conn.

I took a 10-15 second look and turned around and said: "This is a Soviet ****** class oceanographic research ship. Look at about page 270 in that blue recognition book over there on the radar scope."

The operations officer took another look via the periscope as the crew turned to page 271 of the book I had indicated. He turned around, looked at the book, and then at me.

"I'll be damned! You are right again. How do you do this?" I think I modestly answered something to the effect, "It's my profession, and I guess it's called training, sir." It also did not hurt my credibility with the crew at all. I had the advantage of knowing that the Soviets had one of those ships operating in our vicinity, and I had made a point of looking it up a day or two earlier, just to refresh my memory. And I do remember page numbers very well, even to this day. As the storied Alabama football coach Bear Bryant is famous for having said: "You ever notice that it is the well-prepared team that seems to have all the luck?"

One other time I was asked to look at two helicopters onboard a Soviet Space Support Ship. We had received a tasking message to shadow this ship as it moved into position to support a Soviet missile test. Higher authority wanted to know if the helicopters had any unusual antennas on them, top or bottom. The weather was foul, and we could not get a good photo because our lens kept being washed over, so CAPT Cooper wanted me to see if I could see anything. There was a storm on the surface, and as we were flung around, I peered through the periscope to see what I could see. I also clung to it as if it were my life preserver. I finally convinced myself that I could see no differences. It was pretty much the "B" variant of that particular helicopter as near as I could see.

Then, just as I was preparing to hand the periscope back to

CAPT Cooper, we were hit by a wave so big it pushed us far over. Looking through the periscope as I was at that moment only magnified the effect. Also, knocked clean off my feet, I was clinging to the periscope for dear life. I honestly thought we had rolled completely over, which is death to a submarine because the ballast tanks are open at the bottom. One roll beyond 65 degrees or so and done. Indeed, the night before, I had had a nightmare of precisely this happening, except in my nightmare, I watched a movie with the crew on the mess deck. Yes, I was scared.

Then we rolled back the other way, my feet hit the deck again, and I jerked my head away from the eyepiece. I looked around to see the comforting sight of my shipmates clinging to anything they could so they would not be thrown into anything. My heart was racing, but that was all.

My one failure was a bit embarrassing. CAPT Cooper asked me to identify a ship he had been watching, and I looked at it for well over a minute, totally puzzled. It was bow-on, coming right at us, which can make identification difficult. I finally decided it must be a multi-masted sailing vessel that I estimated to be a mile or so away. The Captain took over the periscope again and, after a very short time, ordered the periscope lowered and for us to immediately leave periscope depth (PD), saying: "Take her deep! Fast!" As POGY went into a steep dive, he informed us we had a Whiskey "Canvas Bag" submarine, previously used for radar picket duty, less than 1,000 yards from us, coming straight for us. So much for my infallibility.

WHISKEY "CANVAS BAG" SUBMARINE

That diesel submarine, an early radar picket with an exceptionally large folding antenna, was thought to have been retired. However, there were some reports they were still used as reserve training ships. I wanted to return to PD and take a photo. However, we were much too close, and CAPT Cooper elected to give it a pass rather than risk visual detection by a low-priority vessel. It was the same reason we did not close patrol boats or torpedo retrievers, who appeared after submarine live-fire exercises, which we always covered.

We were on station off Petropavlovsk, the Soviet SLBM base in the Pacific when President Nixon resigned the Presidency. It turned out to be a very scary day. The Sonarmen detected a steady stream of submarines leaving port. They quickly identified the stream to be every SLBM submarine we knew to be operational plus a couple of others came streaming out of the port at best speed in total EMCON. (No electronic emissions at all.) They all immediately went to "Patrol Quiet" and disbursed to scattered positions off the port.

We developed firing solutions, in priority order, on all that we could and sent a burst message to DC giving details of what we were seeing. We also were very attentive to any incoming flash message telling us we were at war. Things gradually calmed down later that day as one after another; the big black monsters slithered back into their holes alongside the piers at Buka Tarya, the submarine base across Avachinskaya Bay from the city of Petropavlovsk. Still, for several hours there, both CAPT Cooper and I wondered if we were in a front-row seat for the start of WWIII. Indeed, we would have been in the center of the ring very quickly. We even loaded at least one anti-submarine rocket just in case we needed it in a real hurry, like when one of these monsters opens its missile hatches. It was very sobering, indeed.

One other event deserves mentioning not for its immediate importance but for several different reasons. Late one evening, we observed a very unusual event. It was so unusual no one onboard had any real idea what it was. We speculated it had to be some sort of covert operation, at least in part because of what appeared to be efforts to conceal what was being done. There were minimal communications, in a remote area, with an unusual mix of vessels, including a Yankee class submarine. Indeed, following the Yankee was how we stumbled upon the event. We detected it shortly after it left the harbor. We thought it might be deployed on an eastern Pacific missile patrol because of the limited communications surrounding its departure, just like a deploying submarine. There was also a hydrographic research vessel with a large crane and some patrol boats. All in all, very odd.

NavSecGru direct support division officers are trained and instructed not to include speculation in the patrol report, so I did not, but I did document what we had observed as accurately as possible in the mission report. Even after the event had secured and all had returned to normal as the ships returned to Petropavlovsk, I could not help speculating about what we had seen. And, even more importantly, what it might mean.

I decided to list my ideas about what it might mean, from most significant to least, and bounced them off the team, which further refined them. All agreed my #1 was #1. I then showed my list to CAPT Cooper, who agreed we had seen something very odd and liked our list. I offered to write a letter from either myself or him to the ComSubPac Intelligence Officer and attach our speculations list.

CAPT Cooper agreed that was the best course of action, and I drafted the letter. In the end, I signed it as a "heads-up" from one intelligence professional to another, not as an official letter. I did this with Captain Cooper's full knowledge and support.

The Commander, Submarine Forces, Pacific (ComSubPac) Intelligence Officer met us at the end of the patrol, did a quick review of the patrol report, and, after asking a few questions, pronounced himself highly satisfied. He congratulated us on a highly successful mission. Then I produced the letter, marked with the highest classification I had available. I explained this was all speculation. However, my team, CAPT Cooper, and I felt it was important enough to call attention to the possible significance of what we had seen and give our best guesses immediately after the event. I ended by saying I left it to his judgment as to how to proceed. He read the letter with our speculations and agreed we had seen something significant, but he had no ideas to add to ours. He promised to send it forward with our trip report, and he would make sure that the right folks saw the letter.

About two years later, after I had returned to Atsugi as the Officer-in-Charge, one of the officers I had broken into the business in 1975 called me on the secure phone to tell me to get over to Naval Communications Station (NavComSta) Kamiseya as soon as I had a spare minute. Kamiseya is about five miles away and has access to higher classified communications than Atsugi. He wanted me to read a very highly classified message from the CIA that he had just forwarded to Kamiseya. My #1 speculation appeared to be right on, and I just might be in line for a high award indeed. I, of course, hustled the five miles over to Kamiseya the next day and read the message.

It went into a lot more detail than I had, but the bottom line was that the letter I had sent and the information we had collected on POGY had alerted the CIA to a grave new threat. It was a good thing I went to Kamiseya that first day because the day after that, I got a call from their Special Security Officer (SSO). He ordered me to come back over to Kamiseya to sign a non-disclosure oath. He also told me to forget I had ever seen that CIA report. It had been reclassified to beyond anything for which I had clearance. So much for my award! Sigh!

However, the story does not end there. In early 1987 I was assigned to an interagency space-warfare working group meeting at the Space Command, Colorado Springs, Colorado. Representatives of all the services and intelligence agencies, NASA, and others were at the meeting. I somehow fell in with the CIA and NSA representatives and a few others from the "Spook community."

That night eight of us decided to go out to a local steakhouse. During the dinner, I happened to use the old saw "You can get a lot done if you don't care who gets the credit." The CIA rep agreed but observed that "It still smarts when you spend nearly two years researching something in your spare time with no support at all from your boss who thinks you are off on a wild goose chase. You finally finish your paper, which was instigated not within the CIA but rather by a letter from a Pacific fleet submarine. It becomes a scorching hot property overnight. Whereon my boss, who had completely discounted my work until that moment, republished it <u>as his work</u> at a higher classification. He eventually even got a large cash award. For two years of my work! I never got a thing. Not even a "Thank you" from my boss, the SOB. I worked on that from the Fall of 1974 to October 1976, but it still rankles to this day, ten and a half years later."

I looked at him and started to say something but decided a steakhouse in Colorado Springs was not the right place. So, I just said I had a question for him when we got somewhere appropriate. I must have piqued his interest because as we walked to the cars an hour or so afterward, he called me aside and asked what my question was.

All I said was: "Cluster xxxxx?" This was what I still remembered to be the name assigned by the CIA to the research paper I had read that day in October at Kamiseya). It was like I had hit him.

"How did you know that?"

"Because I wrote that original letter in the Fall of 1974 on USS POGY. And I saw your original report the day before it was reclassified in October of 1976."

My recitation of those dates and facts immediately established my bona fides, and he congratulated me warmly. We became good friends. It is one of my proudest moments, and I have not told this story more than once or twice in the first 30 years since then due to classification implications. It was that hot!

I think it has now, 45 years later, cooled off enough I can include it here.

As we retired from our patrol area and sailed for Yokosuka, Japan, CAPT Cooper asked me if I was willing to stay on board for his entire western Pacific deployment. He was scheduled to go right back out to the same area and did not want to break up the team. I wanted to go, but there was a rotation system at Misawa, and it was not my decision. I told him he would need to contact my Commanding Officer, CAPT Pete Dillingham, (later RADM, Commander, NavSecGru), and to please stress to him that he had asked me to stay without my suggesting it. He did, and I went home for about a week and then right back to Yokosuka and the POGY.

YAK-28P FIREBAR

The second mission was far calmer than the first. Weather played a significant role in the slow-down, and maybe the Soviet Navy ran out of money to buy fuel for its ships. But, for whatever reason, they just were not coming out to play. We did cover one air defense exercise, and it was interesting. I saw my first ADGM-30, a Soviet ship-based multi-barreled 30 MM cannon used primarily as an anti-aircraft gun. It fired at a target sleeve being towed by a Yak-28P, an early model Soviet light bomber subsequently used as an interceptor and then a trainer. The tracer rounds traced a coil of black rope flecked in red and orange nearly straight up and weaved it back and forth in the sky as it searched for and tried to hit the target sleeve. It reminded me of a cobra swaying to and fro, ready to strike. Thinking of my aviator friends, I shivered. It looked evil and very lethal.

We also saw preparations for land-based surface-to-air missile launches. However, as we moved into position to monitor the event, we hit an uncharted seamount. We dented the lower part of the sonar dome, so we withdrew. We did hear the surface-to-air missiles hit the sea and explode as they completed their run. Not sure why they did not arm and explode in the sky. Exciting day but hitting that uncharted seamount may well have been what cost Dave Cooper his star. It was the Navy's loss. He would have been an outstanding Admiral. I felt terrible because I had become the acting operations officer after the original OpsO had "lost it" and been confined to his stateroom, and I had been the one that had suggested we move into that position to cover the developing exercise better. Sh... happens, even to the good guys!

The most memorable event of my entire time in submarines occurred not long before while we were well offshore observing a multi-submarine exercise. There were three of them chasing each other, and then they started to fire practice torpedoes at each other. We had routinely been cycling between periscope depth (PD) and just below, just as the Soviet submarines were, to catch all aspects of the exercise via SIGINT and sonar. We had just left PD when sonar reported one of the submarines had launched a torpedo, and it had no bearing drift...meaning it was headed right at us!

CAPT Cooper turned to me and asked: "Any indications this is a war shot?"

"None, Sir. I do not think the Soviets would be willing to start WWIII because they discovered a foreign submarine off their coast well into international waters. No submarine commander would take it on himself to launch a war shot without asking permission from his chain of command ashore. They just would not do it, and we have not detected any unusual comms suggesting they have detected us, much less that anyone out here with us might have asked for permission to launch a war shot." was my immediate reply.

"OK. We are just going to sit here and wait this out."

And announced to the crew: "SLOW TO BEAR STEERAGEWAY! COMBAT QUIET!"

And we waited as the Soviet torpedo continued its course straight for us, getting louder with every second. You could have cut the tension with a knife, as they say. Finally, after what seemed like an hour but was something less than 10 minutes, the torpedo went right down our starboard side, close aboard. It was a near miss. A few seconds later, it ran out of fuel and bobbed to the surface, the warhead end up. Someone knew the Soviets paint their dummy torpedo an awful shade of green to make sure fishermen will not steal them, and "our" torpedo was that shade of green. It was a "dummy warhead," sure enough. If they fired it to get our attention, they entirely succeeded. It might also just have been luck that they fired it on our bearing. We never knew. I do know my guys had no indications we were detected, so maybe they were just lucky.

This event is the only event in my entire life where, at its end, I heard grown men giggle like girls. All voices were at least an octave above their normal range! Mine, too, probably. We were all very relieved it had not been a war shot aimed at us. I might have been just a touch more relieved than most because I had made the call, and the Captain had accepted it. I am 99% sure CAPT Cooper and I were the two most relieved men on board that day.

We returned to Guam at the end of the 2nd patrol. My team and I caught a plane for Tokyo and then a Military Airlift Command (MAC) flight from Yokota to Misawa. Thus, ended my operational duty in submarines, but I certainly did not know that at the time. I knew I had done well and expected to make at least two more patrols before transferring out.

Indeed, I was selected to take the USS HALIBUT (SSN-587) out on its 1976 mission. It was "THE" PLUMB assignment for this business by a long way. It was THE SPECIAL MISSION Special Operations boat. LT John Skipper, an up-through-the-ranks guy, led the NSG team on HALIBUT in 1973-1974 and received the Distinguished Service Medal for the 1974 operation. I believe it was the first DSM awarded to a spook for a submarine operation up to that time.

Still, I could be wrong as even the personal awards for those operations were highly classified in and of themselves. I do know several of my friends did get Legions of Merit (LoM) for subsequent special-special (sic) operations and were told they could not wear them until years later. Indeed, that is how I know they got them when, years after the fact, they made Captain and suddenly donned LoMs, which were exceedingly rare in our community. I was going to be his relief in this special-special program, but fate intervened.

First, HALIBUT needed a fair bit of grooming before it would be ready for the 1976 mission. The Navy finally decided to replace the HALIBUT with an even older boat, the USS SWORDFISH (SSN-579), the fourth nuclear submarine ever built. I was not worried; I would take either boat. It was a singular honor just to be assigned to that mission. Then fate stepped in once more.

To say I was proud of my time in submarines is a gross understatement! It was (is?) the pinnacle of US intelligence operations.

NOTE: If you want to know more about US submarine Special Operations in the Cold War, read, in priority order:

1. Red November by W. Craig Reed. 2010 (Parts of it are a near-direct hit on me, but I still cannot, 40+ years later, tell you why!)

2. Stalking the Red Bear by Peter Sasgen (2008, 2017)

3. Rising Tide by Gary E. Weir and Walter J. Boyne 2003 Focused on telling the story's Russian side. Very illuminating!

4. Blind Man's Bluff by Christopher Drew and Susan Sontag, 1998 (The book that broke the story of US Submarines in the Cold War)

5. COLD WAR III by W. Craig Reed (the last 60 pages are especially pertinent.)

Experience Pays Off!
Back to Misawa, to Russia with Love(?), and then to Atsugi

Chapter Eight

My having gone out on two submarine missions back-to-back scrambled the schedule and pushed my next assigned mission five or six months out, which was OK by me. I was a bit exhausted. Not so much physically, but mentally. The tension from the substantial responsibilities of being an officer-in-charge of an intelligence collection detachment on a submarine going intentionally into harm's way is both considerable and very real.

Besides needing a bit of a break, LCDR Bob Carpenter, the new Direct Support (DirSup) division head at Misawa, needed a deputy with significant DirSup experience. With routine transfers, I was now the most experienced DirSup officer at the command.

Indeed, I had completed the "Hat trick!" Participation in operations in all three elements: Surface, Submarine, and Aviation! I do not think I was the first. Still, I cannot recall anyone else doing it during my four and a half years there. I am also quite sure I had more real combat time in this business than any other officer during my time there.

About the time you know your job, you are transferred. It is the US military way. All others with experience, anything like mine in Northeast Asia, had transferred. Also, at about that time, the Navy ran low on Personnel Change of Station (PCS) funds and asked if any of us there in rural Japan were willing to extend for six months to a year. I was doing exactly what I wanted to do, so I readily volunteered to extend a full year.

Bob Carpenter pronounced himself very glad for the experienced help. He also had a new assignee, a LT, about to make LCDR named Daryl Campbell, who had just come into the

NSG. Daryl was a surface officer. He had not gone through the DirSup training pipeline, but his record and obvious intellect had quickly impressed the leadership. (Me, too!)

They had enough operators; the leaders needed managers, someone to ride herd on the operators, such as me. Daryl fit the bill perfectly. Thus, we became a team. He became the DirSup policy thinker, and writer and I became the DirSup team trainer and builder, which was fine for me. Twelve years later, I was still not in the CAPT zone, but Daryl, who was about seven months senior to me when we met at Misawa, had been a CAPTAIN for three years. He was four years in front of his year group! We all thought he was destined to be the next selection for NSG Admiral, but he did not make it his very first time up and retired. I understand this surprised the NSG leadership very much as he undoubtedly would have been selected for admiral the next time around.

Even though I now had a good deal more operational SIGINT experience than Daryl, I learned a lot from him, too. He was very, very smooth and polished, and I certainly needed some of that. We were a good pair for Bob Carpenter, who was also a superb officer. The three of us were a good team supporting our new Operations Officer, CDR Ike Cole, who was yet another star. Indeed, our CO, CAPT Pete Dillingham, and CDR Cole both commanded the Naval Security Group as Rear Admirals. I think the only reason Bob Carpenter did not make it was his health declined. It was an outstanding group, and I was exceedingly proud to be part of it then, and maybe even more so today, looking back.

In the next 18 months, I did go back to Atsugi as a flyer for two more two-month deployments. I went back to sea in September 1975 as "Charger Horse," the SIGINT resource coordinator, and overall coordinator for all intelligence collection of a task group doing "Freedom of Navigation" operations in the Sea of Japan, Sea of Okhotsk, and the Kuril Islands chain. It was a small task group consisting of a destroyer, an oiler, and USS

HORNE (!), my original ship. The aircraft carrier USS MIDWAY and several escorts were also operating north of Japan in the Pacific off the Kuril Islands chain with the support of P-3s and VQ-1 EP-3s from Misawa and Atsugi.

It was a bit sad to see that the HORNE was no longer the bright shiny new penny that I remembered and, indeed, loved. It was only eight years old at that point, but she seemed older. She needed paint and a general cleaning. An intelligence collection space/flag command center had been added on the deck in the area between the two combined smokestacks/masts. The signals exploitation space (SupRad) also has more space, which was good. I did still feel right at home in the Combat Information Center (CIC), even after all the years. Also, it was great to get back to sea, especially in a more senior position!

We sailed from Yokosuka in the late afternoon and, once we cleared the mouth of Tokyo Bay, went straight east for a couple of hours. Then, after we were outside the congested waterways near Tokyo Bay and darkness had fallen, we turned north in total EMCON, arriving off the Tsugaru Strait near midnight. Radiating only our primary navigation radar, a standard commercial system, we transited the strait and entered the Sea of Japan undetected. Increasing speed to at or near flank (as fast as she could go), we headed straight for Vladivostok, home of the Soviet Pacific Fleet. My team and the site at Misawa were listening for any indication the Soviets knew we were coming, but there was none until after dawn when we were sighted by a Soviet picket ship, a destroyer escort, who raised the alarm.

After that, things got very busy for us as numerous air and surface units came racing out to look us over. We turned and went northeast parallel to the Russian coast and then turned and exited the Sea of Japan via the La Perouse Strait, entering the Sea of Okhotsk, basically thumbing our nose at the Soviets.

A great deal happened over the next two days that I will never forget. As we entered the La Perouse Strait, our ELINT team reported we were being illuminated by a Soviet missile boat's fire control radar. We also had a surface contact on the same bearing heading for us at high speed, causing some concern with both the ship's captain and the embarked rear admiral. I frankly doubted that it was a missile boat as there were no missile boats known to be operational in that area at that time. I suspected the radiations were from the Borders Guards Shershen patrol boat known to be stationed in that area. It had a radar that can be mistaken for a missile guidance radar in a search mode. The ELINT team stuck by its analysis until the contact came roaring into view at top speed. We identified it as the Shershen I knew to be assigned to duty in that area. My credibility with the Admiral and his staff took a big boost. I had, after all, flown reconnaissance missions in that area and clearly had more experience there than anyone else on board.

SHERSHEN Patrol/Torpedo BOAT

The next day, while in international waters in the center of the Sea of Okhotsk, two squadrons of 16 TU-16 Badger bombers (total 32 aircraft), each with two air to surface missile slung under their wings, for a total of 64 weapons, launched out of bases north of Vladivostok. They crossed the Sea of Japan and Sakhalin Island and made mock firing runs on us. It was a classic attack profile from both the aircraft flight paths as seen on our radars and our analysis of the SIGINT we detected.

Everyone in Flag Plot was genuinely concerned. We watched the 32 hostile bomber aircraft streamed onto our radar scopes and headed straight for us. Initially, they had just their primary search radars on, turning on their missile fire control radars as the range lessened, then the seekers on the missiles themselves. At about 35 miles out, each bomber broke off to the south, reversed course, and headed home.

TU-16 with Air to Surface Missile

I distinctly remember one officer saying, "I guess they cannot find us and are giving up." after the first few aircraft had flown this profile. It was one of the more ignorant remarks I ever heard in the combat center of a warship. No, they were simulating launching their missiles near the maximum range of our Terrier missile system and flying back out of range as quickly as possible. That day was the 16th of September, 1975. My 32nd birthday. I estimate I was killed in a simulated fashion about 64 times that day.

Our small task group proceeded across the SOO, exiting through a strait in the Kuril Island chain into the Pacific Ocean. The carrier MIDWAY was operating to our southeast, and when we electronically linked to them, I felt a lot better. We also established links to both the P-3 and EP-3 flying in support coverage for us. That made me feel even better.

As we sailed south, opening out from the Kurils, two TU-95 Bear D maritime reconnaissance aircraft, flying out of a base northwest of Vladivostok, closed in on us as they illuminated us with their fire control radars. The Bear D does not carry missiles itself but instead uses a data link to pass fire control information generated by the large radar in its belly dome to other missile-equipped units, including ships, ground sites, and submarines. These two Bear Ds went through the whole launch sequence, just as the TU-16s had done the day before.

TU-95 Bear D (Maritime Reconnaissance Version)

I recognized that to do that, they must have a missile-equipped recipient in the area, and it needed to be in a relatively confined geographic area. We knew there were no Soviet missile-equipped surface ships or ground sites in the area, so they must be linked to a submarine. And that it must be close to a specific Kurile Island strait. We informed the Admiral of our analysis, and he alerted the P-3 in the area. Sure enough, they got a contact close to where we thought it probably was.

We were ecstatic about that part of the exercise. All in all, it was extraordinarily successful, especially since the Soviets had reacted so strongly and given us great collection opportunities.

Truth be told, they probably felt the same about us as we turned on our weapons systems to be ready to defend ourselves if they did launch on us. This was at the Cold War's height, and we never knew when it just might go hot. I realized that we had reacted to their reaction in such a way that the OsNaz, the Soviet SIGINTERs, my counterparts, must have been delighted, too.

They now knew how we would engage many missile-firing aircraft, such as the double squadron of Badgers, who used us for target practice. It was all part of the cat and mouse game of the Cold War. Sometimes one is the cat, and other times, the mouse.

In 1968 North Koreans had shot down a VQ-1 EC-121 in international waters well off its harbor of Wonsan. The US had "rattled its saber," but nothing came of the incident other than the Navy stopped flying reconnaissance missions against North Korea. In 1975 it was decided that the US Navy would resume flying intelligence collection missions against North Korea. I was selected to lead the SIGINT collection portion of the mission. It was a true honor, and I was delighted. It also meant that I needed to go into yet another training program. It was straightforward in some ways as it was the third country program that I had taken. Still, much of the equipment was the same as North Vietnam and North Korea were Soviet client states and used older Russian weapons.

ECHO II SSGN (with missile tubes raised)

It was a bit harder because we did not have that much raw data on North Korea at Misawa or Atsugi. This was overcome in part by the need also to qualify some of our Korean linguists for airborne operations. There were plenty of volunteers, but they were all stationed in South Korea. Our Operations Officer decided to solve these related problems by sending the volunteers from South Korea to SERE/DWEST, just like I had done.

From there, they went to Atsugi to acquaint themselves with airborne operations by flying on Sea of Japan missions. I was also sent back to Atsugi at the same time to be both their instructor and their student. My task was both to educate myself and form the team. These guys were the best Korean linguists in the Navy, and they were very motivated, so it was an absolute pleasure to work with them.

After several weeks at Atsugi, we deployed to Taegu Air Base in the center of South Korea. However, we first visited the USAF watch center at Osan Air Base, not all that far south of Seoul and the Demilitarized Zone (DMZ). We discussed warning procedures. We also wanted to know at what alert status the USAF F-4 Phantoms there at Osan would launch. They were promised to ride "shotgun" for us, and we also wanted to know how long it would take for them to get airborne and join us.

At the first sign of hostile air activity, we were assured that at least two F-4s would be launched. There were four fighters on immediate ready alert and another four on standby. More than two fighters would launch only if the North Koreans launched more than two MiGs in reaction to us. In talking with the fighter squadron guys and learning of their eagerness to engage the North Koreans, we began to feel more like bait on the end of a hook.

We hoped all would go as planned but were a bit concerned over the always-present possibility in any military operation of a SNAFU (Situation Normal, All Fouled Up). My team and I joked among ourselves about this, but it was always met with nervous laughter. Luckily, nothing untoward happened, and we executed our three or four missions without a hitch. It was a learning opportunity for all concerned, VQ-1, USAF, NSG, and me personally.

Joint Mission planning conferences for all reconnaissance flight operations in the western Pacific are held monthly at the 5th Air Force Headquarters at Yokota Air Base. From subsequent discussions at these conferences, I learned that our command chain was glad the Navy had gotten back involved in flying reconnaissance missions against North Korea. No one knew when that volatile nation might reinitiate hostilities, and the US needed to be ready. Indeed, this is still the situation today, 40+ years later.

About two years later, after I had become the Officer in Charge at NSGD Atsugi, the VQ-1 Duty Officer woke me up at 1:30 AM. He told me the National Command Authority (i.e., The White House) had a report from the CIA that North Korea was planning a surprise attack the next morning. We were the only asset immediately available to be on scene at dawn. We were ordered to launch as soon as possible.

The only plane ready that day was an EC-121, the oldest airplane in the squadron. We got airborne in record time and flew straight into a hurricane in the southern Sea of Japan. It was my only hurricane penetration, and many of us thought it would be our first and last. We gyrated all over the sky as the hurricane tossed us at its whim.

I remember a receiver at my position nearly broke free. I reached up and pressed the substantial piece of equipment back into its slot, adding my weight to the restraining straps. I wanted to ensure it would not come completely free and start swinging wildly about the crew compartment. I spent the last 90 minutes of the flight to the DMZ holding it in its place. I know others had similar problems.

I was very glad when the flight engineer re-secured the receiver once out of the hurricane as we arrived on station just south of the DMZ. After several hours, we determined it was a false alarm, much to everyone's, including POTUS's, relief. All in a day's work. I was glad we were called on and had been able to respond. So was the entire chain of command, up to the White House.

The EC-121 was also used for hurricane hunting, but those aircraft are set up for that mission, with special reinforcements inside to hold the sensors and their operators in place. Our airplanes had not gotten that specific treatment and were never meant to go inside a hurricane, but there we were. The ride through that hurricane was something all of us will remember for the rest of our lives.

Back at Misawa, I took my training duties very seriously and studied right alongside my charges. We were more like a seminar study group, working together, helping each other learn more and get better prepared to do what all of us thought was one of the most exciting jobs in the US military. I remember sitting at the big table in the study room with about ten fellow DirSup officers and suddenly realizing many of us had a lot in common besides our Navy careers. Most of us were Catholic-educated and had experience in music. Most of us also found math and physics interesting and easy but were much more interested in its application than as science in and of itself. There were, as you can imagine, some jealousy bubbling under the surface, but we were all gentlemen. I do not remember any of them being jerks, but anyone could see the guys who were going to be the stars or at least thought they were. I know I did.

History pretty much proved me correct, although one guy, Alex Miller, made admiral, which I would not have predicted. Just did not see it. He was smart enough but not aggressive or "hungry" enough; thus, I was not really surprised when he did not make a second star. Still, in the Naval Security Group in those days, promotions went from 50 (?) LTs a year to about 35 LCDRs to 12 to 14 CDRs to 4 or 5 CAPTS. Admiral is selected every two or three years, so making admiral is a very big thing. In that group of a dozen-plus LTs at Misawa in 1975, I picked every one of the guys that made CAPT, which was over half of them. They all deserved it. But none of us took it for granted at all at that time.

Sometime in late 1975, LCDR Bob Carpenter and CDR Ike Cole informed me that I had been selected to be the DirSup team leader for the 1976 deployment of USS HALIBUT (SSN-587).

My clearance was upgraded to a level I did not even know existed. It was to be the second mission to an extremely sensitive area since its upgraded capabilities were added. However, after much discussion well above my paygrade, the Navy decided that HALIBUT was beyond economical repair. The oldest nuclear submarine in service by some years, the USS SWORDFISH (SSN-575), took its place.

After that one trip in 1974, HALIBUT was retired. Its place was eventually taken by PARCHE (SSN-687), the last and newest of the Sturgeon class, possibly the best "Spook boat" ever, but I cannot tell you why…and there may well be even better ones on duty today, ones I know nothing about. The Russians now know enough about these operations to build a small fleet of their own special-special mission submarines. I understand many in western naval intelligence are genuinely concerned about this threat. Still, it has been years since I last had those clearances, so my knowledge today is limited. I can only speculate and wonder.

USS Halibut (SSN-587) with Diving Chamber

To be selected for a special-special mission was a big honor, but I was told not to discuss it with anyone. However, I did go through another security briefing. I was given the combination to a safe in a very private part of our considerable office space on Security Hill at Misawa. In the safe were pertinent historical reports and many other interesting items. As was suggested to me by the leadership, I drove the six miles out to the office from my home in Misawa town at least once a week at night to read these files without anyone around.

I did this for the next several months. During that time, I was told to start selecting my team. There was a set of men already chosen for these operations, but I would need to make my needs and wishes known ASAP if I wanted anyone else.

EASIER SAID THAN DONE! You needed a cross set of skills and personalities that meshed, but I had first pick, and in the end, I went with the same guys that had made the 1974 trip. Each was already briefed into the program and knew what they needed to know for this mission. We were all pumped. But it was not to be. It was a major disappointment in my life, but like every other one, it all seemed to work out for the best in the long run. But it does still rankle just a bit even today.

NavSecGru Detachment Atsugi, Japan
My First (and only) Command
Chapter Nine

We were getting close to final team training for the USS SWORDFISH mission when LCDR Bob Carpenter, my division officer, called me in and said words I hate to this day.

"Guy, no reflection on you at all. And I really hate to do this knowing how hard you have worked to get ready, but you are the only qualified officer even remotely available for in immediate assignment. We need you to leave for NavSecGru Det Atsugi as soon as possible. You must be ready to assume the duties of the officer-in-charge (OIC) there when CAPT Dillingham (our Commanding Officer) relieves the current OIC "for cause." We anticipate that it will be a week or so after you get there.

The Commanding Officer of VQ-1 has reported to CAPT Dillingham that our OIC is just not hacking it as the test director of the Deepwell System. Deepwell was the heart and soul of the NavSecGru manned mission system of Navy's new EP-3E. The first EP-3E was delivered to Atsugi had, not surprisingly, proven to have significant problems. Others were starting to roll out of the factory and be delivered to VQ-1 at Atsugi and VQ-2's detachment in Athens. Deepwell was an automated collection system, the first attempt at automating this task anywhere in the world. As far as pure computer capability, it was a step too far, but that was not clear at that time.

NSGD Atsugi was charged with conducting the acceptance test and certifying it was ready for fleet operations for those planes delivered to VQ-1. A few of the operators had been part of the initial training and acceptance test cadre at the factory, but most of the operators at Atsugi were expected to learn this new, highly complicated system via on-the-job training. However, all this was further complicated because the senior operators clearly felt threatened by the system. The older operators felt challenged by the automation. They believed the computers were trying to replace them and were only too happy to shut off the Deepwell system and revert to the manual mode at the first indication of trouble.

The biggest problem was that they were not documenting what they were doing when it stopped working. There was an apparent failure of leadership at the OIC level, probably because he had no clue what his fundamental role was. He was also up against many other challenges that he just was not equipped to handle, including the revolt of his senior non-commissioned officers, something which is exceedingly rare in the military, but something I saw again when I went to the USAF and was put in a very similar situation almost three years later with the Rivet Joint test team.

I think either our operations officer, CDR Ike Cole, or his deputy, LCDR Postom, maybe both, had flown down to pay a visit to Atsugi. They were both very unimpressed. I cannot remember if it was one or both of those gentlemen who visited Atsugi, but I think it was both because the day after I was told I was going to Atsugi on the next flight, they both had me into their offices, serially, to tell me what I needed to do very quickly to regain control of the detachment and the test effort. The test effort was in addition to its regular several days a week reconnaissance missions, so I knew I would be remarkably busy.

I immediately packed my bags and headed down to Atsugi for the most challenging four months of my life. This was doubly tough because this is the same time I would have been engaged in the special- special (sic) mission, for which I had spent several months in secret training. That still rankles more than a bit.

After I arrived at Atsugi and reported back what I saw to be the situation, they decided not to have me relieve the OIC. On my recommendation, he was allowed to serve out his last four months of his tour and leave at his regular rotation date. However, he clearly understood the implications of me being pulled out of the special-special mission team and coming down to Atsugi very nearly unannounced. My arrival was wholly off schedule, and the rotation plans, officer and enlisted, were typically made and announced months in advance. I was supposed to be getting ready to go on an unspecified submarine special-special operation which had just been announced then I suddenly arrive at Atsugi? Everyone knew why I was there. To say the relationship was strained is putting it mildly.

One other item of significant contention at Atsugi was caused by the fact that the designers of the EP-3E had omitted any data systems collection and analysis capability in the system. This was something that both the VQ-1 leadership and I thought was a huge mistake, given the evidence we were seeing that our potential opponents were "going digital."

EP-3E

The detachment had also been stripped of its three billets for men with qualifications to sit such a position, the Cryptologic Technician Technical branch personnel with "Special Signal" or "non-Morse search and development" training. They went by the nickname "T-Birds." They were second-class citizens when they had been onboard the EP-3B and the EC-121s. The linguists had a very high opinion of themselves. They believed they were the only ones of any significance on the aircraft. I subsequently discovered this was also true in the Air Force. Indeed, they were critical to mission success, but mission success took a team. Think of a football team made up of just quarterbacks facing a well-balanced team. Any fan knows who would win that game!

As I mentioned in the submarine operations chapter, the difference in the signal environment between my two tours in Vietnam, plus my submarine special operations experience, led me to believe data signals would become critical in modern warfare. We needed to be ready for it. Thus, I supported the idea of bringing T-Birds back to the detachment. It was a very unpopular idea there at Atsugi. It was met with open skepticism by the senior NCOs, all of whom were linguists.

There were only so many collection positions in the airplane, and putting a T-Bird at one would mean we would lose a linguist position. After I became OIC, I suggested the linguists train themselves to operate the required "T-bird" equipment. You would have thought I was asking them to forsake their birthright or have sex with their mothers. The submarine SIGINT training pipeline had included training on the T-Bird position. I had taught myself how to be a passable T-Bird, and in that, all my men were clearly at least very smart if not brilliant, I knew it was possible.

Indeed, my experience in Vietnam and Northeast Asia over the previous eight years convinced me that data links were becoming more important, even critical, in tactical and operational roles, especially in the tactical world. Our fundamental, primary mission was not to collect intelligence as most of the detachment believed, but rather to be ready to support combat operations at any time.

As I detailed earlier in this story, in October 1973, Nixon threatened to strike the Soviet Union if they used their paratroopers in Israel. That day I thought we might go to war with less than eight hours warning. Now that I was the OIC, I wanted my men to understand the necessity to be ready to go to war with no notice. Being able to exploit datalinks just might be the margin of victory. (That was true even then, but it is especially true today!)

The guys eventually understood my reasoning, and things

leveled out, but the first several months were very tough. We ultimately got T- Birds reassigned to the Detachment. In our sister detachment in the Mediterranean, the T-Bird is one of the busiest men on the crew, but the eastern Med has always had a very sophisticated signal environment. I am sure that is true even today.

Besides being ready for war at a moment's notice, we also had to document what was wrong with the Deepwell system. With the help of the Deepwell technical representative (TechRep), I spent a week making myself as knowledgeable as possible on the problems with Deepwell. I also asked for his ideas on how to fix the problem. My time in the NTDS room on HORNE and my several computer science classes paid off handsomely here as I at least understood what he was saying.

EP-3E Aries II

Once I understood what path to start down, I met with the senior NCOs and appealed to their professionalism. We may have been handed a lemon, but we owed it to the Navy, the nation, and ourselves, to document what was wrong and why. With their skeptical but thoroughly professional help, the senior NCOs and I developed procedures to document what was happening when the system froze, which it often did. We also documented how we were able to restore it to service, which we seemed to be able to do about half the time. It was a learning experience for all of us. The TechRep was delighted with my initiative, but it was, in truth, not mine. It was straight out of one of my computer course textbooks.

Two or three years after I left Atsugi, the Navy leadership decided to scrape Deepwell based on our reports and those of our counterparts with VQ-2 in the Mediterranean. Trying to use a 15K computer (upgraded to 64K about a year into my tenure) to run multiple positions (6/7) in a tactical support aircraft probably seems just ludicrous to anyone in the service today. Still, we just did not understand the problem in those days, and the engineers believed that the installed computer was all that needed back in the mid-1970s. Even as late as 1982, when I took the "Intro to Computer Systems" for the third time, people just did not understand how easy it was to overload the early computers.

I distinctly remember my instructor, the head of IT for a "Big Eight" accounting firm suggesting in September of 1982 that any of us contemplating buying a home computer hold off for six months. The Japanese had just announced they had developed a 64k CPU, which was four times the size of most CPUs on the market at that time. It was going to be available in less than six months, in the spring of 1983, "and 64K is all that anyone will ever need," according to my instructor. And he was one of the leading lights in the business computer world at that time. Tell this to a computer professional today, and then stand by for a lot of laughter!

As I mentioned previously, Chuck Christman, "The Crab," is world-famous as the most innovative man in ever airborne reconnaissance. He and I had a very cordial relationship, more so than any other NSG officer at the time, maybe ever. The Crab was always working in some critical area, constantly pushing the technical ball forward. He had an idea for a signal processor. His partner, Emler Achtenberg, also mentioned earlier, was able to develop the math to make it work. Chuck called his latest invention the Markham Automatic Readout Box (MARB). I was absolutely fascinated by it. I spent a lot of time talking with him and my team about how his tool could be used in combat operations. I was the only NSG officer who initially showed any significant interest in the MARB, probably because I immediately understood why Chuck's "black box" would be both important and necessary.

Chuck needed funding to perfect the box. He asked me to recommend his work to CAPT Dillingham, which I was glad to do. We were hoping to get funding from the Navy's NavSecGru side, or maybe even the National Security Agency. CAPT Dillingham officially went to NSA to ask for help. We did get technical support, but not the money needed to buy just the materials to make the required parts. The Crab also went through his VQ-1 chain of command but to no avail on either front. It was very frustrating for me, but doubly so for Chuck.

I met Chuck four years earlier, in September 1972. I was a NavSecGru flight officer just assigned to fly with VQ-1. Prior to that, in 1968, I had spent five months off Vietnam on USS HORNE (DLG-30), where I had learned how to use the WLR-1 electronic intelligence receiver. I returned to Vietnam in April 1972 and immediately noticed that the spectrum held a much-expanded set of radars and new data signals. Always curious, I became highly interested in the exploitation of both the radars and especially the data signals.

Chuck had been flying SIGINT reconnaissance missions in Vietnam during the entire war. He had noticed the same thing and was working to understand the use(s) of the data signals and how to read them, so we had a lot in common. We became good friends as I helped him with his research. I became one of his many "disciples." I do not recall any other NavSecGru officer doing that at the time. They were, to a man, focused on the foreign-language aspects of our duties. The VQ-1 naval flight officers were trained to identify and collect radar signals. Chuck and I, plus a few others, were looking at the data signals. We had a lot to talk about as the signal environment seemed to change weekly. That was one reason why I believed we needed our CTT "T-Birds" back on the plane.

His effort finally bore some fruit, but not because of anything I did. In the Fall of 1974, Chuck attended a PARPRO conference at Yokota AFB. He met Senior Master Sergeant Barry Mathews, an E-8 USAF sergeant from the Headquarters, USAF Security Service, San Antonio. They fell to talking about data signals and their potential usefulness in combat. Sergeant Mathews mentioned that the USAF had come to the same conclusion. It was the #1 intelligence collection and analysis priority in the Strategic Air Command.

Chuck told him about his magic box, now sitting on a shelf at Atsugi gathering dust. Atsugi was about 25 miles from Yokota. Sergeant Mathews was interested, so Chuck offered to give him a full demonstration. Sergeant Mathews called his boss, a colonel at San Antonio, and asked for permission to extend his stay in Japan for two days. He wanted to go to Atsugi to check out Chuck's box, the MARB. The Colonel said, "You are a big boy. If you think it is that important, go for it." (I would be assigned to that job in San Antonio four years later.)

Chuck and SMSGT Mathews drove to Atsugi that night. The next day, Chuck got the box down off the shelf, dusted it off, and hooked it up to a recorder. The demo clearly showed the potential to Sergeant Mathews. He asked if he could borrow the box and ship it to the USAF Big Safari Detachment at E-Systems at Greenville, Texas so that they could study it. They were the Air Force's research and development organization focused on airborne reconnaissance. Chuck had run out of ideas on how to get funding to continue its development, so he readily agreed.

When the MARB box got to Greenville, the smart folks at E-Systems and USAF Big Safari Detachment 8 were very impressed. They made a copy and sent the original back to Atsugi, where it was returned to its place on the upper shelf in the storage shed.

When I became the OIC at Atsugi in June 1976, I learned that the USAF had E-Systems developing a tool based on the

MARB, called the EPR-107, but we saw nothing in the field.

This next episode figures exceptionally large in the rest of my professional life, but I am getting well ahead of myself.

6 September 1976 is one of those dates that will stick in my memory forever. It was mid-afternoon on a hot and humid Labor Day (just like DC) in Atsugi, Japan. I was with my family at our cramped off-base house with no air conditioning when I got a very unusual phone call. It was rare we got a call at all. Receiving one in the middle of the afternoon on a holiday had never happened before, so I sensed the call was special even before lifting the receiver. Even more unexpected, it was from Chuck Christman, "THE CRAB," himself. He did not engage in idle chitchat, so I knew it was a significant call even before he said a word. It was most definitely not going to be a social call.

Just as I anticipated, his first few words electrified me. "A Foxbat has landed in Japan!" Chuck had just gotten a call from the duty officer at the US Embassy in Tokyo. A top-secret MiG-25, our #1 collection priority at that time, had just landed at Hakodate, a civilian airport in southern Hokkaido, 120 miles north of Misawa Air Base, about 600 miles north of Atsugi. It appeared that a defector had flown it in from a Soviet airbase near Vladivostok. We subsequently learned that the pilot was Lt Vicktor Ivanovic Belenko, Soviet Air Force.

The rest of this story is of the Navy's role in examining and exploiting that Russian MiG-25. It all starts and ends with Chuck Christman. Known as the most innovative person in Navy airborne reconnaissance, he was always hard at work in some critical areas, pushing the technical ball forward, expanding "the edge." The USAF had its Big Safari, but we had "The Crab."

Chuck knew I had contact with a highly classified US government office near Atsugi whose mission was liaison to Japan's government on military technology matters. He asked me to call them.

Chuck was hoping to use my connections at that unit to get us assigned to the exploitation team. Immediately after Chuck's call, I called my contact at the technical liaison organization. Telling him that a MiG-25 had just landed in northern Japan, I asked if Chuck and I could meet with him the next day to discuss how we could help each other capitalize on this event. My friend was most interested and invited us to visit him. Early the following day, Chuck, his assistant, Stuart Jefferies (another very bright guy), and I drove to his highly classified office.

After introductory discussions, we offered any assistance we could provide and, among other things, described the capabilities of the MARB. They immediately became highly interested in the idea of testing the MiG's electronics with it.

The organization's leader made a series of calls on a secure line to the headquarters, US Forces, Japan (CUSFJ). Within minutes he was talking to the Commander himself, a US Air Force Lieutenant General. Our team was immediately given the top priority for exploiting the MiG and ordered to take the MARB and our test equipment to Yokota Air Base as soon as possible. Military aircraft would take us north to the site as quickly as possible. Chuck and Stu were elated. They were going to get to test their system in as real a setting as could be imagined. I was delighted to be part of a team with such an exciting task.

We immediately returned to Atsugi and started building the test stand, but the next 24 hours were the rainiest of my entire life. We had 15" of rain in 12 hours and then over 10" in one hour, followed by it tapering off to an inch an hour for the next three hours.

All in all, it rained 28" in less than 20 hours. Even the runway was utterly awash for most of the afternoon after the rain had stopped. Even the runway was at least two feet deep in water with waves. The golf course, which we could see out the back door of my detachment's workspaces, looked like a raging sea as mountains of water cascading down the hills. Work came to a standstill but resumed with a vengeance as the roads cleared and we could get to the VQ-1 electrons shop where we were assembling the test equipment.

The following day we moved our equipment and ourselves to Yokota AFB, the home base of US Forces Japan, and the 5th Air Force, for further transportation to Hokkaido and the Foxbat, but were told to standby in the BOQ.

Five days later, we were still waiting to go north to the test site. I think the Japanese government did not wish to acknowledge they had asked Americans for assistance for two reasons.

The first reason was the political one. Japan wanted to bargain with the Soviets, with the idea that this was big enough the Soviets might be willing to open negotiations on the Habomai Island group. These are a set of islands at the southern end of the Kuril Islands chain just to the northeast of Hokkaido, the northernmost of Japan's four major islands. The Soviets had seized the Habomai Islands at the end of World War II. The Japanese considered these islands part of their homeland and wanted them back.

Secondly, they also wanted to save face because they did not want to admit they had no test equipment anywhere near as good as what we brought to the game.

I was in the 5th AF watch center writing a report to my bosses in Misawa, CAPT Dillingham, and CDR Ike Cole, both of whom went on the become rear Admirals and led the NavSecGru when the USAF watch officer came over to my table and asked for a favor. He had an off-line encrypted message for Prime Minister Migi from our President, Jerry Ford. He could not leave the watch center, and in that, I was the ranking military member of the exploitation team in the area at that time; late in the evening, would I please hand carry it to the US Ambassador? He was visiting the 5th AF Commanding General in his office, a floor above. "Of course, I'd be delighted!"

It was in a folder which was not sealed, and he did not tell me not to read it, so my curiosity got the better of me, and I read it before I delivered it to the Ambassador in the Commanding General's office — absorbing, indeed! Apart from some of the off-line encrypted messages for the CNO that I saw as a LTJG while part of his communications guard, probably the most interesting letter I ever read. We were prepared to give quite a lot to get our hands on a nearly brand-new MiG-25.

A day or two later, the Soviet Foreign Minister landed in Tokyo to bargain for his airplane. The press asked him if he was willing to discuss the trade of the Habomi's in exchange for the MiG-25? His reply was classic Russian. "The Habomi's are an integral part of the Soviet Union. There is nothing to discuss." Period. Suddenly, both the Japanese Prime Minister and the Foreign Secretary were too busy to see him. The Soviet foreign minister soon returned to Russia empty-handed.

Once he left, we were given the green light to go north with our equipment. Chuck and Stu went to the site, but I was diverted to a sequestered Japanese barracks at Misawa. I could not contact anyone, but I was kept very busy answering questions from the Japanese in any case.

Soon after I got to Misawa, CDR Denny Wisely (who went on to be a Rear Admiral), a LT whose name escapes me, and Major "Fergie" Ferguson, USAF, arrived at Misawa. They all had experience flying various MiG aircraft and were there to fly the Foxbat if anyone could. CDR Wisely was judged to be the most experienced pilot in similar aircraft, so he was to get first crack at flying the Foxbat, followed by Major Ferguson and the Navy LT.

About a dozen other Americans, all US civil servants, also arrived from several US organizations. The USAF's Foreign Technology Division had several engineers, each an expert in an aspect of aeronautical engineering: engines, airframe, metallurgy, avionics, etc. Their home organization had given each specific task, and naturally, each believed theirs was the most critical task.

However, the US on-scene commander was Colonel. Keith Carnahan, USAF, the FTD Detachment Four Commander, based at Yokota Air Base. His office was near the Commander, USFJ, and he worked directly for the General. Colonel Carnahan had his instructions that the MARB tests were first priority. Nothing was to interfere with that task. There was more than a bit of grumbling as Chuck and Stu were among the very few allowed to have unlimited access to the MiG.

MiG-25 FOXBAT

Our Japanese hosts warmly welcomed Chuck because he, having lived in Japan for over 20 years, and having a Japanese wife, was fluent in Japanese. Also, they quickly understood how important his work was. They also welcomed me as I was conversant in Japanese, having studied it for the last four and a half years, since arriving in Japan.

When they discovered I could also read Russian, they started bringing me photos of the cockpit and its instruments. Because so many of the questions were redundant, I started making drawings of the instrument panels, left, right and center. Using a Russian technical dictionary that I had brought with me at my chiefs' suggestion, I began labeling each dial and lever's function. The Japanese liked that a great deal and brought in a draftsman to make my amateur drawings into a finished product to brief their Japanese seniors and others in the Japanese government, giving me full credit. I was a popular guy. I have no idea why Japanese intelligence did not assign several of their Russian linguists, which I am sure they have, to work with me, but they did not.

About nine days later, the MiG was loaded onto a USAF C-5, the free world's largest airplane at the time, and flown to Hyakuri Air Base, Japan's military aerospace test center. It had two distinct advantages as a test site. It was both much more secluded than Hakodate's civilian airport and near Tokyo. I, with most of the team, returned to Yokota. I was getting a bit nervous about my impending Inspector General (IG) inspection, so I asked Colonel Carnahan, the on-scene commander in charge of the overall US effort, to release me to return to my detachment to prepare for my IG inspection. I offered to send him CTI1 Tom Botulinski, my best linguist, in exchange. He thought I was a good enough linguist, but I assured him truthfully that Tom was much, much better. Indeed, Tom was one of the very top linguists in the Navy. It was a good deal for all involved, including Tom.

The Colonel was very much a gentleman. IG inspections in all services, especially the Air Force, are particularly important, so he understood my situation perfectly. He thanked me very sincerely for my efforts and for being a team player. Some folks were not! They were trying to smuggle out information so their home organizations could "scoop" the other organizations there. This was causing him trouble both with his boss in Japan, ComUSFJ, the USAF general, and his boss at FTD Wright-Patterson AFB.

We parted on the best of terms, and he subsequently nominated me for the Meritorious Service Medal, but it was downgraded to the Joint Service Commendation Medal. I was just happy to have been there.

Tom Botulinski took my place and did a bang-up job for them and even got to go up and work on the plane itself, something I never got to do. I subsequently urged him to put in for the Limited Duty Officer (LDO) program. He did and retired many years later as a captain. His last job was Deputy Director, Naval Security Group, second in command of NSG. As I said above, Tom was exceptional!

I needed to do one other thing before I was finished. The MiG had run off the end of the runway upon landing and, its nose wheel tire was destroyed. The Japanese aircraft mechanics checking out the Foxbat had determined that the MiG's nose wheel was the same size as the main mount of an A-4. However, neither Japan nor the USAF fly the A-4, so they had no tires for it. As the ranking Navy guy from Japan, they turned to me for help. So, I asked the Navy aircraft maintenance shop at Atsugi to lend the team a tire. Once the exploitation event was over and it was decided that the MiG would not be test flown, after all, the Navy wanted their tire back. So, I flew back to Yokota AFB in a Navy chopper to facilitate the wheel's return.

As we were loading the wheel into the helicopter (it was a tight fit), the Air Force Senior NCO of Det 4, who I had been dealing with constantly from the beginning, came to attention. He saluted me, saying something like it had been an absolute pleasure to work with me and that I was a credit to the US's armed forces, even if I was a Navy man. That meant a lot to me at the time and still does. You do not have to salute on the ramp of an airfield, so he was paying me a huge compliment. I had seen our effort as a team effort from day one. Others had been more out for their command and themselves, and it showed. By working with the Colonel and his team, rather than trying to go around them, I had made their job easier, not more challenging and they clearly appreciated it.

They also appreciated Chuck and Stu's work with the MARB, which was the main effort I was recording and reporting. Chuck and Stu were doing all the "real" work. I was just the scribe. But everyone was delighted with all our efforts. The MARB performed as hoped, and we now knew we had a useful exploitation tool.

E-Systems subsequently sold a number of their EPR-107 processors, the device they developed by reverse-engineering the MARB, to the US Navy for a substantial amount of money.

I got back to my detachment, and the senior NCOs there had the situation well in hand. I was not surprised. We passed the IG with only very minor suggestions on how we could improve our operation. CAPT Pete Dillingham, the CO at NSGA Misawa, pronounced himself very pleased, and if he was pleased, we were all delighted.

His fitness reports on me went a long way to getting me selected to attend the Naval War College. That had been a dream of mine since before I joined the Navy in 1965. But I never thought it would happen because the competition is so fierce in the NavSecGru. At the LCDR level, 112 men were considered for one slot. To be selected means you are still in the hunt for promotion to Admiral.

I had some genuinely exceptional men working for me at Atsugi. I encouraged many of them to apply for either Warrant Officer or Limited Duty Officer (LDO), citing myself as an example. I started as a deck force seaman, a boatswain's mate striker, the lowest of the low. I was now Officer-in-Charge (OIC) of one of the two premier airborne intelligence collection detachments in the US Navy. Clearly, the opportunities are there if you reach for them.

Near the end of my tour, the Navy Times published a short article about how our 35-man detachment in Japan had seven of its men picked up for officer programs in the space of just over a year. I had no idea it was coming, but we were all delighted.

I now know the Public Affairs Officer at NSGA Misawa wrote it. According to the article, which I knew nothing about until I saw it in print, this was a record percentage of both application and acceptance. It is a fact I am proud of to this day, but it was the men themselves who stood out and made it all possible. All I did was point out the way and the rewards.

There were many other things, both operationally and administratively, that I was proud of at Atsugi. Two other things stand out to this day.

First, my counterpart OIC of the detachment in the Mediterranean, working with VQ-2, and I collaborated to get Naval Air Observer (NAO) wings authorized for the NavSecGru officers who flew on the VQ missions. Before that time, only meteorologists who flew ice reconnaissance missions were permitted to wear NAO wings. If you deleted the words "ice" and "meteorologist" from that section of the Navy Uniform Regulations, we were already qualified.

Still, the Navy made us jump through all sorts of administrative hoops and set significant requirements before they would authorize the wings for NSG officers. We asked for and received endorsements from both VQ squadrons and the 6th and 7th Fleet commanders plus the area fleet commanders, Pacific Fleet and Atlantic Fleet. There was a lot of pro and con discussion about this on both sides of the globe. Ultimately, we were successful, and the NAO wings were authorized as of 1 July 1978.

My change of charge ceremony was scheduled for 20 June, ten days before the wings were authorized to be awarded. Both CAPT Tettlebach and his Executive Officer, CDR Stuart Huntington, had helped get the NAO wings approved. Wanting to personally award me the wings we had all worked on getting authorized, they went back to the Navy chain of command, asking for permission to grant me the wings a few days early. Thus they could hold the "pinning" ceremony for me, and I would report to my duty station with the USAF wearing them. So, it was done.

Therefore, I was the first NavSecGru officer to be awarded NAO wings, notwithstanding the Navy Times' story written by someone in the Mediterranean claiming that my counterpart and his boss to be the first two NSG officers to be awarded them.

The article was correct in saying that they had been a big part of the push to get the wings, and I do not care who was first, but it was nice, right, and just that the men who had helped me get the wings authorized had presented them to me. Also, the fact that I arrived at the USAF command wearing wings was all I cared about, and it was to pay huge dividends.

The other administrative initiative that I instigated was an even tougher fight. LT Dick O'Neil was THE workhorse of my detachment. He flew many more missions than anyone else at Atsugi, either NSG or VQ, and his work was always top drawer. He also helped run the processing and reporting side of the detachment's business, which was directly related to flight operations. His advice was always spot on. He was so valuable, and his work was so unique, that I decided to make him a test case for the Air Medal award for the peacetime aerial reconnaissance program (PARPRO), as our flight operations were called.

I enlisted help from both the VQ and NSG chains of command in Japan and the 7th Fleet and Pacific Fleet staff cryptologists. We got Dick his Air Medal, the first awarded to anyone for PARPRO operations. It was richly deserved, and I am proud to have led the effort that got it done.

About ten years later, Dick was awarded the best student paper prize at the Naval War College. I was on the staff there then and happened to run into him that evening as I was leaving the Newport Officers Club. He and his wife were just arriving for a celebratory dinner.

His first words to his wife, who I had not met until then, were: "Honey, this is the man who taught me how to write. I owe him many thanks." And, after I thought back, I understood why he said that. His first reports at Atsugi had been all over the place, and I had given him the advice given to me many years before. "Tell them what you are going to tell them, tell them what they need to know, and then tell them what you have just told them. When you finish your report, go back and place your summary sentence at the beginning of the report."

Dick, another very bright fellow, quickly became a master of the written word and retired as a remarkably successful Captain.

He now owns a very highly regarded think tank.

The first real indication I was doing better than most of my contemporaries was when I came up for my orders after Atsugi. This was three years before I was selected for the Naval War College, my second strong indication. The detailer called to tell me he wanted to send me to Pearl Harbor as the 3rd Fleet Cryptologist. Because the job was a commander's billet, and I was a brand-new lieutenant commander at the time, I thought he was offering me the deputy's position. I told him I would have to think about it. I had been promised an exchange job with the Air Force's equivalent to the NSG, the USAF Security Service (USAFSS), in San Antonio if I would extend my tour in Japan for six months. I had extended for the requested time, so I was a bit put off that they were reneging on their promise.

I called CAPT Ted Tettlebach, who had replaced CAPT Pete Dillingham, and expressed my frustration. He phoned the detailer, and I got my promised orders to the Air Force in San Antonio. However, as he ended his call telling me he had been successful, his parting words ring in my ears to this day.

"Guy, you must really want San Antonio. That job as the Fleet Cryptologist is a big step toward the flag, especially since they were willing to give it to you when you were a rank junior." Say, what? Shot myself in the foot, there! I thought they were offering me the Deputy Fleet Cryptologist job. It is an honor to be offered a job at a rank senior to your level in any position, anywhere, any time. And, as well, I was brand new in that junior rank.

Thus ended my operational life with the USN. It had been one hell of a ride! Thank you, God! I could not have done all this without a lot of help!

At the end of July 1978, I reported to the United States Air Force Security Service (USAFSS) headquarters, Kelly Air Force Base, ready to start to "hang up my spurs" and be a paper pusher. Boy, was I in for a surprise!

US Air Force Security Service
My Career Takes Off –
Literally and Figuratively

Chapter Ten

    I arrived at USAF Security Service Headquarters, San Antonio, in late July 1978, wearing my new Naval Aviation Observer wings and a well-trimmed, very red beard. I was assigned as the head of the Line-of-Sight Collection division, Current Operations Directorate, the same job the previous NavSecGru exchange officer had. That division was responsible for all USAF SIGINT collection manning and readiness in the line-of-sight regime, both airborne and "remote."

    This included "Space," which was so classified in those days that it was not even referred to outside of highly classified areas. We also had tertiary tasking authority, behind the Operational commander and the Intelligence Community, up our chain of command. That included all aircraft, ground stations, and satellites. My two bosses, Colonel Bobby Berry and his assistant Colonel Grover McMakin were both highly experienced intelligence officers. My officemate was CAPT Walter "Rosy" Rosenstrom, an up-through-the-ranks legend in the USAF airborne reconnaissance business. The job also included managing the readiness of all linguists in the USAF.

    On day three in my new office, I arrived to find a copy of USAF Uniform Regulations, which included grooming standards, on my desk. There was a paperclip on the page about facial hair. I read that paragraph and turned to Rosy. "Who put this here?" "Colonel Mac himself." came the reply.

    "I think he wants me to shave my beard!"

    "You are pretty smart for a Navy guy!"

I thought about the fact that I was a naval officer, and my beard was within Naval Regulations. I also thought about the fact that my job was to get along with my Air Force hosts and that it would probably be counterproductive to engage in a test of wills over something so trivial in the big scheme of things. So, even though I loved my beard, and many others did, too, I went home that night and shaved it off. I had not shaved since that fateful night on the USS WORDEN, back in April 1972, when my not shaving had probably saved my life, but it was the right decision. I had looked like a Viking with the beard, but now I appeared very ordinary to myself, and I suspect to everyone else, too.

That was probably exactly what the USAF desired.

The next morning Colonel Mac saw me as I walked in the door and greeted me like a long-lost son. Later that morning, Colonel Berry had me into his office and complimented me on my new, clean shaved face. More importantly, he told me that he and Colonel Mac had read my award citation forwarded to them by NSGA Misawa to present to me. They now understood my role in bringing the EP-3E to life and were impressed. Given that experience, they had discussed my highly unique qualifications in getting the EP-3E online and my SIGINT combat experience with the Commanding General.

All three had agreed it would be an excellent idea to put me on USAF flight orders and visit all half dozen USAF reconnaissance bases: Offutt, Mildenhall, Athens, Okinawa, Shemya, Elmendorf. I would fly a few missions from each base to cement my understanding of USAF reconnaissance flight operations in hostile areas. They thought this would take 6 to 9 months. At the end, I would come back to the headquarters and write a proposal on how the USAF and the USN could do a better job to coordinate and collaborate in the reconnaissance business. This task would also introduce me to the entire USAF SIGINT reconnaissance operation. This experience would be beneficial for the other job they had in mind for me.

At that time, the USAF was building the Rivet Joint Block III (the RC-135W), the second aircraft in the world to use computers to run its primary mission system. (The E-3E being the first). The General and the Colonels all thought it would also be a good idea to have me help test that airplane and they immediately assigned as the deputy test director for its operational evaluation (OpEval). I would start helping to write the highly detailed test plan that day. What did I think about this idea?

My reaction was immediate and extreme. I replied that I did not think I had the imagination to dream up such a perfect set of orders! Less than a week later, I was enroute to Offutt Air Force Base, Omaha, Nebraska, where I flew a pair of missions on the RC-135V Block II testbed. It was flying against a Navy task force in the Gulf of Mexico, and they thought I might be able to help them beat the Navy. It was all great fun, and I was very impressed with both the professionalism and size, both in breadth and depth, of the USAF reconnaissance operation. It is almost a side-show in the Navy by comparison. Not so in the Air Force, where it is a "front-row" operation.

I was immediately sent to the Air Force's equivalent of SERE school, SV-83, near Spokane, Washington. It was already near freezing there. Quite a climate change from Texas. Its focus was on how to survive in a prison camp. This was different from the Navy's focus, which was more on evading capture, surviving, and being rescued, but those subjects were covered there, too.

We spent several days in a mock prison camp. The instructors tried very hard to break us via sleep deprivation and abuse. I do not remember being waterboarded there as I had been at Navy SERE school, but I suspect I was. Having been to Navy SERE/DWEST and then participated in many refresher training days with VQ-1, this training was a piece of cake. It probably also helped that I was now an O-4, and we understand the Russian prison camps respect rank, so I think they gave me a little slack, not that I asked for it at all. In the end, I passed and was now qualified to fly in hostile areas with the USAF, and fly I did.

Less than a month later, my first deployment was to Okinawa, flying right back to the western Pacific, where I had flown with VQ-1. This helped me a good deal as I was entirely familiar with every aspect of the mission and understood every position in the RC-135M (Rivet Card) in which we were flying. It was similar to the EP-3B that VQ-1 flew, except the EP-3B had a much better ELINT suite and less than half as many COMINT positions. The RC-135M was primarily a COMINT collector, while the Navy flew SIGINT (COMINT & ELINT) collectors.

My first mission was to the Gulf of Tonkin, where our primary collection target was Vietnam. As we entered the area and moved to our tasked mission track, our operators noted some fighter activity on the Vietnamese air bases at Da Nang and Vinh. Da Nang had been a US base, and Vinh is a significant Vietnamese base on the central North Vietnam coast. As we turned north to commence our mission's first leg, the chatter on the airplane's intercommunications system grew as operator after operator reported potentially hostile activity. Wow! This is getting exciting! We seemed to have stirred up a bit of a hornet's nest, or was this a game of "Scare the Squid?" Were my new AF crewmates testing me to see how a Navy guy might react in a potentially hostile situation?

RC-135M RIVET CARD

I looked around me, and every operator I could see, about a dozen, seemed genuinely concerned. They were putting on their helmets and flight gloves and securing loose items in their positions. Was this all a show for me? I thought about this for less than a minute and decided these guys were apprehensive for real. There were probably not many good actors in all the Navy, and I suspected this was also true in the Air Force.

As we flew north up our track, we were moving into a dangerous situation. If the four MiG-21s preparing to take-off from Da Nang had instructions to engage us, they would be in an excellent position to do so. On launch, they would be between us and where we would need to exit the Gulf of Tonkin. We did have the option of overflying China, but that was not a real option. As tension built and remembering my accident over the Sea of Japan (SOJ) after my EC-121 went into a steep dive, I stopped walking around the positions looking over the crews' shoulders. I had stepped through a hole trying to get to my ditching station that day in the SOJ. I still walk with a limp because of it, so I quickly headed back to my ditching station against the mission compartment's rear bulkhead. I was extremely glad I did. Events proved in the next few minutes that I had done a very smart thing.

All the operators had put their helmets on and checked their survival gear by this time. I had not been belted in but about a minute when the Airborne Mission Supervisor (AMS), a senior sergeant, declared a Condition One ("Hostile intent demonstrated"). It was the first and only Condition One I ever experienced. The pilot responded, "Roger the Condition ONE!" and threw the airplane into an immediate, violent, very hard, very abrupt, sharply descending right turn.

Very quickly, we were pointed almost straight down in what most would think was a Boeing 707! The sudden, violent maneuver caused everything that was not completely secured to go weightless and airborne. This included clipboards, pencils, pens, and most worrisome of all, coffee cups, which 2 seconds before had been resting peacefully in their holders at each position.

After that moment of weightlessness, the aircraft accelerated from something like 375 MPH to almost 500 MPH as the big jet started into its dive in earnest. All the items that had been suspended in the air a moment before came flying back to my position, pelting me and the wall around me. Some of the coffee cups were hitting the wall so hard they were exploding, and the thought raced through my mind that life really was unfair. After surviving many hostile situations, I was going to be killed ignominiously by shrapnel from an exploding coffee cup!

Luckily, I survived that barrage, too.

The pilot extended our landing gear to slow our descent to a controllable level, making the airflow over the aircraft very turbulent in this nose-down position. It felt as if we were falling down a long set of stairs. We had been at 30,000 feet when we started our dive. The pilot told us later that he had tried to level us out at 500 feet, but even after we were wings level, nose up with full power back on, we continued to mush on down to about 200 feet. This was the second time in my career that I had pulled out of a dive lower than intended. Neither time was much fun!

We climbed back to 500 feet and headed away from Vietnam. As we exited the Gulf of Tonkin at a low level, we went right over some of the contested islands in the South China Sea. The Raven, the electronic warfare officer, a major, was on his career's last flight. His ditching station was next to mine, and he looked as if he was having a heart attack but finally regained his color and composure. He had never been this low over the water before, and he got out of his ditching station to look out one of our few windows, portholes, in reality.

He was cautious to keep his oxygen bottle with him. I had to ask why the bottle? "SAC regulations! Anytime you leave your position on a mission, you must take your oxygen bottle with you!" came the answer. That made me laugh, which broke the tension. As I laughed, he suddenly understood the silliness of carrying an oxygen bottle around when you are at 500 feet. He had spent his entire career operating above 25,000 feet, where oxygen just might be needed if you had a sudden loss of pressure, but down at near sea level, where we were now, there is simply no need. He suddenly realized why I was laughing, and he laughed, too. We certainly did need that comic relief just then.

We landed at Kadena AFB, Okinawa, Japan, without further incident, but the aircraft immediately went into a rigorous inspection, checking for overstress on the airframe. (There was none. Boeing "ironworks" does indeed build strong airplanes!)

Thus ended my first operational flight in the US Air Force. Just like my other "firsts," this had been a memorable occasion. Over the next two and a half years, I spent significant time flying on many different reconnaissance tracks (flight routes) and became good friends with many USAF officers and enlisted troops.

Two of the airmen, in particular, stand out in my memory, Sergeants Riley Perdue and Skip DeRouse. They were both masters at their craft and very willing to talk about it to anyone with the clearance who would listen. I listened and learned. They appreciated the audience and the questions, which meant that I was listening and trying to learn. We remained friends throughout our careers and beyond. Skip was a cowboy from Texas in his heart and recently passed away after a phenomenally successful career as a contractor with a defense firm in Austin. I miss his irreverent and salty counsel. Riley is more than a bit smoother and has gone on to have a super career in the intelligence world. Indeed, I worked FOR Riley in the Navy's TENCAP (Tactical Exploitation of National Capabilities) office in the late 1990s as we brought the Integrated Broadcast System (IBS) online.

Three months after 9/11, he was a very senior person at a 3-letter agency and tried to recruit me. By that time, I was deeply immersed in replying to the Presidential task to develop what became Maritime Domain Awareness and turned him down. However, he was the first person in the intel community to which I described my idea for satellite AIS, asking for NSA's blessing, which he helped get for me. More on that in the last chapter of this book.

I also flew reconnaissance (spy) missions out of Mildenhall, England; Eielson, Alaska; and Athens, Greece, as well as Offutt, Nebraska. But my plans to also fly out of Shemya, Alaska, were derailed when our detachment at Athens sent in a message to the headquarters. It had a long list of questions about how Block III would help rather than hinder them. And what was USAFSS going to do to get them ready for a completely new system and capability? I was tasked with answering their many questions, and I turned to my senior sergeants for their input.

As I discussed how we were going to address Athens' concerns with my group of highly experienced NCOs, it became evident that many of the concerns expressed were remarkably similar to the situation that I had seen at Atsugi. There the senior non-commissioned officers (NCO) had initially tried to kill the Deepwell system. The NCOs at Athens were trying to get ahead of the game and list their concerns before they had ever even seen the Block III. I liked their attitude better. My experience with the EP-3E and the Navy NCOs helped me know what to say to the USAF's senior NCOs. In the end, my boss liked our draft response so much that he ordered me to go to Athens to personally reassure them that the new automated Rivet Joint Block III would not degrade their efficiency but rather enhance it.

I had to cancel my flight to Shemya, which disappointed me, but my visit to Athens went well. And, the flight I would have been on to Shemya ran into a wind shear on final approach and crash-landed extremely hard, tail first, just short of the runway. All six men in the rear section of the plane, where I would have been sitting, were crushed to death. I had ducked another one!

Less than a month after I returned to San Antonio after my

first deployment to Okinawa, the US Army briefed the Director of the CIA, Admiral Stansfield Turner, my old commanding officer on USS HORNE, and the man responsible for me becoming an officer, that the space systems controlled by the CIA and used by the Army in their Tactical Exploitation of National Capabilities (TENCAP) program, would be both more effective and cheaper than either the Navy's EP-3 or the Air Force's RC-135 fleets in the reconnaissance role.

Admiral Turner was intrigued by the idea. He even got his picture on Time Magazine's cover with several satellites flying over his head and a caption of "Mission Impossible." One of the article's main thrusts was that the CIA Director was looking to Space to improve intelligence collection and save money. He froze the Navy and USAF airborne programs' budgets, asking both services to re-justify their airborne reconnaissance fleets given the Army brief.

The heads of both Naval and Air Force intelligence, Rear Admiral (RADM) Shapiro and Major General (MGEN) Marks, and the Director of the National Security Agency (NSA), Vice Admiral (VADM) Bobby Inman agreed to put together a team to write the re-justification. It was, in essence, a rebuttal to the Army's claim.

As the acceptance test director of the mission system of the latest version of the EP-3, and the only person who had flown on both Navy and Air Force SIGINT reconnaissance aircraft in recent memory, I was a natural to be added to the team. However, no one knew of my personal relationship with Admiral Turner at that time, and I decided not to tell anyone for several reasons. But mostly, I wanted my ideas and arguments to fly or fail on their own. I did not want to be used to get a specific point of view from Admiral Turner.

There were seven mid-grade officers and senior NCOs assigned to the task, including my officemate, Captain Rosy Rosenstrom. We met at NSA for four days (Monday through

Thursday) to start our mission. I arrived on Sunday night.

In a letter just after he had arrived at the CIA, Admiral Turner had asked me to call him the next time I was in DC. He also asked me to wish him luck in his new assignment, saying, "all I know about the intelligence career field I got from following your career." This is the first time I have been back to DC since then, so I decided to phone him at home the night I arrived as it was about 8 PM on a Sunday. As gracious as ever, he said he was delighted to get my call and invited me to have lunch with him at CIA Headquarters on Tuesday. I, of course, immediately accepted, even though I had no real idea what my schedule was going to be for that week.

As it turned out, the first day for the team was a series of in-depth briefs on our national intelligence space systems' abilities. What they could do today (as of October 1978) and where their capabilities were projected to go in the next 5, 10, 15, and 20 years. It was fascinating. The following day the Navy and the Air Force gave briefs on their manned reconnaissance aircraft over the same time. It was apparent to all of us that while there was some overlap between space and the airborne world, there were also substantial differences, especially in a tactical role. So, we started writing the rebuttal late morning on Tuesday and kept at it on Wednesday. I made several contributions to the draft, but at about 11:00 AM on Tuesday, I excused myself, apologizing that I had an important meeting with a very senior officer that I could not attend. Absolutely true!

I drove from NSA to CIA Headquarters. Admiral Turner's aide met me in the lobby and escorted me to the Director's office via a private elevator. The lunch went along fine until the Admiral asked me exactly what I was doing in DC. With that direct question, I figured I had the best level with him, and I did. He was glad I was part of the effort to answer his question and asked many questions (he always did). I tried my best to answer them, but some of his questions were on things we had not even addressed in our initial draft there at NSA, and I told him so. I assured him I would return to the team and would be sure we answered those questions.

I arrived back at NSA mid-afternoon as the team was about to wind down. I introduced the questions as items we should consider the next morning. When the following day came, I reiterated we needed to be sure these questions were answered. Some of them were so basic that the team felt they were a waste of time. I reminded them that Admiral Turner was not an intelligence professional but rather a Navy surface officer. While highly intelligent, the actual intelligence system was new to him. We might just need to establish the basis for our answers. That won the day, and we included answers to the questions the Admiral had asked me.

I had succeeded in keeping the fact that I was relaying questions directly from the Admiral. However, Captain Rosy Rosenstrom, my office mate from Texas, detected that I was substantially more forceful in my insistence on the inclusion of several new points. Indeed, I had not even mentioned them before I had left the team for lunch "elsewhere." That night, over a beer, he asked me point blank, "Who in the world did you have lunch with? Admiral Turner's aide?" Close, but no cigar!

I decided I had best explain to him that I knew Admiral Turner personally, and he was the source of my questions. He spat his beer all over the bar when I told him I had met with Admiral Turner one-on-one. After he got over his complete shock, I asked him not to let the rest of the team, and he promised not to tell the team. The next day he fully supported my questions as a core part of the brief.

We finished up the paper in the early afternoon Thursday, and the team disbursed. Rosy went to catch an early evening flight to San Antonio. I drove down to NavSecGru HQ in northwest DC to confer with my Navy colleagues. Always a good idea to keep in touch with your home base. I now know that before Rosy left for San Antonio, he phoned our mutual boss,

Colonel Bobby Berry, and reported my conversation with Admiral Turner.

The first I knew of this was when I got a phone call at NavSecGru HQ from VADM Inman's office at NSA. "You are to immediately phone Colonel Berry at San Antonio and keep calling until you talk to him. Also, you have an appointment with Vice Admiral Inman at 0715 tomorrow morning here at NSA." On the other end of the line was Admiral Inman's military assistant, CDR John Thomas, the NavSecGru DirSup officer on HORNE back in the Fall of 1968. The man who has suggested I might like a career in the NavSecGru.

However, he professed as not to know why Admiral Inman wanted to see me first thing the next morning. I could only assume the worst. I was sure he wanted to have my head on a spike at the entrance to NSA by noon after I had been "drawn and quartered." The old traditional English punishment for traitors.

I finally got through to Colonel Berry, our division officer. He wanted a first-hand account of my discussions with Admiral Turner. He was a warm and kind man, and after I described my conversation with Admiral Turner, he assured me I was in no trouble with my Air Force office. Indeed, he was flying up on a very early flight the following day. I was to meet him in Colonel Jim Clapper's office in Air Staff offices at the Pentagon.

Jim Clapper was the same officer who had extensively debriefed me on the exploitation of the MiG-25 in Japan in 1976. He had spent nearly the whole day asking me questions. At that time, he had been deep selected for Lieutenant colonel and was waiting for promotion. He subsequently spent the absolute minimum time as a Lieutenant Colonel. Now, just over two years later, he was a full colonel in an especially important billet in Air Force Intelligence HQ. He went on to a spectacular career, commanding the Defense Intelligence Agency, creating the National Geospatial-Intelligence Agency, and ending his career as the Director, National Intelligence, THE top job in US intelligence.

Colonel Berry also indicated we had an appointment with MGen Marks, the head of AF Intelligence, late that morning, but he wanted to talk with me first. My only possible reply was, "Yes, Sir!"

I immediately changed my return trip to San Antonio from mid-Friday afternoon to Saturday morning. I booked another night in the Howard Johnson Motel there in Laurel. It beat a Navy BOQ of those days, but not by much. It was not as good as any AF BOQ. I was ever in, but the military got a special rate there. It was near NSA, so all USAF personnel visiting NSA were directed to stay there.

I probably could have just sat in my car all that night as I was genuinely concerned about what my reception would be in VADM Inman's office and did not sleep a wink. I met him when he was a captain and the N-2 (Naval Intelligence) at Pacific Fleet Headquarters. He visited Atsugi in 1976, not long after I became Officer in Charge. I had also interfaced with him as the Director, Naval Intelligence. I knew him to be a brilliant man, and he seemed fair and straightforward. Still, he and Admiral Turner were known to not be on the same page on many issues. Also, VADM Inman was said to be more than a bit frustrated with a non-professional intelligence officer as the Director, CIA, so I could only guess what my reception at his office was going to be.

I need not have worried. I arrived at VADM Inman's office a few minutes early and was ushered right in by CDR John Thomas. The Admiral met me with his disarming smile and asked me in an amiable tone to describe my meeting with Admiral Turner, and I did. The Admiral asked questions for about 20 minutes and then got up from his desk, apologizing for cutting the meeting short, but he had a weekly staff meeting to attend in the conference room down the hall. I had been in the Admiral's office for about 30 minutes at that point and did not think it was a short meeting at all!

As we walked out of his office, he put his arm around my shoulder! I was more than a bit relieved and a bit shocked. He indicated that he was delighted that I had been able to make some of the same points he had been trying to make to Admiral Turner. Based on Admiral Turner's questions, I seem to have made him aware of the need for the real-time, tactical intelligence capabilities that only aircraft can give you. That information would be critical in wartime. Satellites, on the other hand, might be better in providing strategic, long-term intelligence. Based on the direction of his questions during my time at the CIA HQ, VADM Inman believed I might have gotten this through to ADM Turner.

Thinking back on my discussion with Admiral Turner and the essence of his concerns that was the thrust of his questions, I understood precisely why VADM Inman had reached that conclusion, but only after I thought about it for some time, but VADM Inman had picked right up on it. He was a very bright guy. As we left his office, he warmly thanked me again in front of his entire office staff, including CDR John Thomas. I singled out John to the Admiral as the man who had introduced me to naval intelligence and suggested I make it a career. Had I been in any trouble, I would not have embarrassed John that way.

I left NSA feeling totally pumped up, assisted by the fact I had had a dozen or so cups of coffee to compensate for the zero sleep the night before. I drove down to the Pentagon to keep my appointment with Colonels Berry and Clapper. I found the right office after a reasonably short search. Not an easy task, as the offices for USAF intelligence were buried in the basement. Colonel Berry arrived shortly after that, and he and Jim Clapper proceeded to grill me on my now-famous meeting with the DCI. At the end of that session, Colonel Berry suggested I stress some things and omit others in our next appointment, which was with MGen Marks, head of USAF Intelligence.

That meeting also went very well. I left the Pentagon feeling incredibly happy but still a bit nervous about where all this was going. I caught a flight back to San Antonio the following morning.

I was settling down to a typical weekend when I got a call from Colonel Berry early Sunday evening. Did I believe Admiral Turner would be willing to tour a Rivet Joint if we could get one into Andrews Air Force Base, just outside of DC, the same place Air Force One is kept on Ready Alert? I was sure the Admiral would be most interested in touring one if it could fit his schedule. He was aware he did not know much about the Rivet Joint, and I was sure he wanted to learn more.

Colonel Berry's next question absolutely got my attention. The Chief of Staff, USAF, the most senior officer in the USAF, had already authorized a Rivet Joint flight to Andrews AFB for this brief. Would I be willing to phone Admiral Turner and invite him to tour the airplane and determine when the tour could be scheduled? Again, my only possible reply was, "Yes, Sir. I will do it immediately." And I did.

It was almost exactly seven days to the minute since I had called the Admiral at his home with such eventful results, so I believed it was probably a good time to phone him, so I did.

I told him I was forced to describe my meeting and his concerns to both a member of my team there in DC and then to VADM Inman and MGen Marks, as well as my boss in San Antonio. His first question was, "Are you in any trouble?" I had to reply that that had been my first thought, too, but now it seems that all were delighted that he and I had had our discussion, and I was instructed to invite him to tour a Rivet Joint at Andrews if it could fit into his schedule. He liked the idea a great deal and gave me the name and number of his aide. I was to call the aide Monday morning. He would already be aware of the call's purpose, and we were to work out the details.

I phoned his aide at the requested time the following morning, and he and I set up a meeting in the time window that worked for all concerned. I immediately reported the same to

Colonel Berry. He asked me to work with the Air Staff in the Pentagon, specifically Jim Clapper, to prepare the brief and arrange the details.

The date was nearly a month in the future. The chief briefer, Brigadier General (BGEN) Doyle Larson, the father of the RC-135 fleet, and now the J-2 (Intelligence Officer) at Commander, US Forces Pacific, was going to fly in from Hawaii. He was the USAF's expert on airborne reconnaissance. I would also have a briefing role. We were to do a "Huntly-Brinkley," with the General describing USAF operations and me doing the same for the Navy. Being paired with a general of another service is not a bad gig for a Navy LCDR!

I coordinated the brief with the Air Staff and with BGen Larson's office in Hawaii as best we could, given we did not have email in those days. I also let my contacts at NavSecGru HQ know what I was doing and going to say, but they were not invited to the brief, even though I had tried to get them an invitation. However, I did not have the horsepower then to make that happen. This briefing was going to be an all-Air Force show, except for Admiral Turner and me. All the services: Army, Air Force, Navy, and Marines, play rough ball all the time, so it was no surprise to anyone on either team, Air Force or Navy.

In the end, BGen Larson and I rehearsed the brief in the morning before the afternoon brief to Admiral Turner and three of his staff. BGen Larson obviously knew his "stuff" and was a very forceful personality, so I intentionally took a back seat, but my few injections were welcomed, especially by Admiral Turner, who made a point of explicitly asking me questions. I think this was more to demonstrate to the assembled USAF brass that we actually did know each other and that he trusted my opinion, which I much appreciated. He was that sort of a very thoughtful leader.

Over the next 30 months, every time I saw VADM Inman, he asked me if I knew Admiral Turner's position on various topics. Even if I did not know specifically, which I often did not, Admiral Inman was interested in my best estimate as to what it might be. He was a great leader and a very gracious man, too.

At the end of the brief and the Rivet Joint walk-through, Admiral Turner indicated he was extremely impressed. He now had a much better understanding of its considerable capabilities, many of which he now understood were unique. He went back to his office and released the funds for the operation and upgrading of both the RC-135 and the EP-3 fleets.

I was the hero of the hour, if not the year. It was very heady stuff. Colonel Berry told me, jokingly, I am sure, that I did not need to come into the office for the rest of the year, except to pick up my paycheck. (I had my pay sent directly to my bank, but he did not know that, but I appreciated the thought.)

Shortly after that, the USAF gave me the Air Force Commendation Medal for my part in "saving the RC-135 fleet." It was all a bit embarrassing, considering what I had done to earn my two Navy Commendation Medals. Still, I did wear the Air Force medal.

I went back to flying out of the rest of the AF operational bases I had missed in my first pass. Within the next three months, I visited every USAF base that operated reconnaissance aircraft that shared hostile airspace with Navy reconnaissance aircraft. I had spent exactly 100 of my first 300 days with the AF on official travel. I had been writing my report as I went along, so it did not take me long to finalize it after I had finished my tour of the USAF bases.

As I was finishing up my flight tour of the AF bases, the USAFSS commander was relieved by newly promoted Major General (MGEN) Doyle Larson, the man I had met at Andrews AFB during the Admiral Turner brief. We had a short but very significant discussion at his welcome aboard reception. He pronounced himself very glad that I was on his staff. He also had done his homework. He knew that I had participated in the special submarine programs, been involved with USS HALIBUT operations, and knew of Cluster xxx, a classified Navy program.

He also told me he had visited a nuclear submarine at Pearl Harbor just before its departure on its special mission patrol. The NavSecGru team had briefed him on board. He had come away very impressed and believed that those teams were the best-trained SIGINT collectors in the US military, cross-trained in all SIGINT disciplines. He compared the submarine team leaders with the Airborne Mission Supervisor (AMS) on a Rivet Joint. He frankly admitted that the submarine team was better trained across a broader range of disciplines. I, of course, did not disagree with him. Indeed, I knew it to be true from personal experience. Still, it was not something I ever mentioned to anyone in the USAF, but I was impressed the general acknowledged it!

The Navy guys were not more intelligent; it was just that we went to an excellent set of schools before we were let anywhere near a submarine. We had to know every position cold. That entailed four different skill sets in the SIGINT world. Even today, 40 years later, I am reluctant to define those specific skill sets due to some of their sensitivity. We also had to know the submarines' communications capability. All that schooling at Ft Meade and Groton paid huge dividends once we got in the fleet (and on beyond.)

General Larson subsequently read my report and its recommendations on how the USAF and the USN reconnaissance efforts could better collaborate, including my proposal to exchange flight crew members for a month or two at a time. He liked the idea, but many of the senior NCOs in the command did not. They could not believe the average Navy operator could be as competent as they were. They were sure I was some sort of a "ringer," and there were very few with my level of experience in the Navy. I kept telling them I was somewhere back in the middle of the pack, and there were many guys in the Navy better qualified than I at any one specific position, which, I am sure, was correct.

I did have two unusual strengths for a linguist. Those were my interest in data communications and my knowledge of radar systems. I saw data systems as the wave of the future, as did MGEN Larson. Another plus for me! There were not that many of us that believed this in 1979, but time has certainly proven us right.

I was more broadly trained than anyone I knew at all seven different positions in the RJ. Still, I also knew Navy guys that were better at any one position than I was. Several years later, I met CAPT Don East, an ex-enlisted NSG linguist in FOUR languages, as well as a Morse intercept operator, and a T-Bird in both RADAR and SPECIAL SIGNALS. He had become an officer and was the immediate past commanding officer of VQ-2 at the time. I am sure he was better than I at most, if not all, positions in the RJ and EP-3. I also suspect he was not alone, but I never met anyone else. I would have loved to introduce him to my Air Force colleagues. He would have "watered their eyes!"

I was in an interesting position at USAFSS HQ. The previous general had been an F-4 fighter pilot, and he had staffed the HQ with his fighter pilot colleagues. Few, if any, had any real experience with SIGINT. Colonels McMakin and Berry were two of the only four colonels at the HQ I could identify with actual signals intelligence experience. There was a like number of lieutenant colonels and maybe a few more majors.

I remember Colonel Jerry Wuscher, one of the other two colonels with an intelligence background, remarking that the number of qualified intelligence officers with operational experience at USAFSS HQ at that time would not fill up a small bus. I have no doubt he was right.

Indeed, General Larson put out a memo that said literally, "Amateur hour is over." That might even have been its title. In the announcement, he stressed that he wanted all his majors and above to be fully conversant with SIGINT. If they were not interested in doing this, they should immediately apply for a transfer or retire. Wow! As one can imagine, that note was the best-read document most of us had ever seen and a subject of much discussion around the HQ.

Not long after I submitted my report on USAF/USN airborne collaboration, General Larson and I went to DC to brief it and hold discussions on it with pertinent Air Force and Navy offices. We went via a T-39 VIP transport. There was just the general, his aide, Major Bill Bucholtz, who I had worked with in Japan during my previous tour, and myself. Bill was busy doing paperwork for the general, but the general engaged me in a wide-ranging conversation on many subjects starting with how to implement the suggestions in my paper and moving on to world affairs.

I distinctly remember him saying that while there were many hotspots in the world, two were on the edge of going extremely hot, just waiting to boil over. One was the collection of countries known as Yugoslavia. Its leader, Tito, was near death. With him gone, General Larson believed Yugoslavia was going to fall into warring factions. The other was Afghanistan. It was falling apart, too, and the Soviets just might invade it soon. We could end up deeply embroiled in either or both places. We had better be ready. He made these remarks in late November 1979. At the time, I had never heard any similar forecast, and I read the daily highly classified intelligence reports religiously. History very quickly proved him correct. I guess that is why he was a general in intelligence. He knew his stuff.

We gave our briefs to both the Air Staff and Navy elements in the Pentagon and then went to NSG HQ at 3801 Nebraska Ave NW, Washington, DC. That complex is now the Headquarters of the Department of Homeland Security (DHS). At all places, people expressed interest and support but not wild enthusiasm. Still, the general asked me to work with all these organizations to develop an unofficial aircrew member exchange program as a step toward better coordination and collaboration. Mission accomplished. New mission assigned.

The general flew back to San Antonio, but I took a day of leave and stayed behind to see my brother Bob, then a commander in the US Navy's Supply Corps. He had recently transferred from Hawaii to the Pentagon. I had not seen him in 2+ years, and it was good to see him. When I arrived at his home, I learned there was a birthday party for him on Sunday night, and he suggested I take another day off and stay over one more night. So, early Saturday evening, I called Colonel McMakin, Colonel Berry's deputy, and asked for permission to stay another day. His answer electrified me! "Yes, but be sure you are in the office bright and early on Tuesday. General Larson has just appointed you as the Director of the Rivet Joint Operational Evaluation (OpEval) and Acceptance Team. You two must have had one hell of a conversation as you were flying to DC! You are now 0-4 in a senior O-6 position."

The colonel who had led the RJ OpEval was a SAC B-52 pilot. He and the general had gotten crossways over something that must have been personal as I never learned what it was. He was fired on Friday, immediately after MGEN Larson returned from our trip. On Saturday morning, the general called a meeting of his six most trusted colonels, the four intelligence professionals I had identified plus an acquisition expert and a maintenance specialist. I subsequently got to know all of them very well. The topic of the meeting was who would replace the just-fired colonel.

Colonel McMackin told me that, after a bit of discussion and no obvious answer, he spoke up, saying: "General, I hate to admit it, but easily the most qualified man here is our Navy LCDR." The general readily agreed, and I was immediately assigned to one of the most significant colonels' assignments in the USAF, much less the USAF Security Service, to be the acceptance test director of a brand spanking new airplane! Wow! I found it hard to believe it was happening, but it did.

I spent the next 27 months in that job. Of those 27 months, I spent all but three deployed. First, to E-Systems, Greenville,

Texas, where the Rivet Joints and many other interesting airplanes were, and still are, created. (The most famous two aircraft made there were Air Force ONE, the president's plane, and the National Emergency Airborne Command Post (NEACAP), the most powerful airborne communications system in the world.)

Once we had certified the aircraft ready for mission operations at Greenville, we took it to Offutt AFB for crew training and further testing. Finally, we deployed to Mildenhall AFB for more crew training and intensive testing in the most arduous hostile signal environment in the world at that time, the Baltic. I had the responsibility and authority, with many strings, over budgets of $249M one year and $332M the next. It was a heady experience.

For the first 18 months or so, I kept thinking that any day I would be summoned to General's offices where I would be introduced to a squared away USAF colonel who looked like he had come straight out of central casting. The conversation was going to go something like: "Guy, you have done a good job, and this is absolutely no reflection on you, but we need to have an Air Force officer lead this effort for appearance's sake if nothing else. Meet the new Rivet Joint Test Director. You will be his deputy, and I know you will give him your full support with your technical knowledge as the total professional you are." That conversation never happened, and I must admit I was, and am, somewhat surprised it did not.

I had a test team of 74 USAF aircrew men. They were the "pick of the litter" for every skill onboard the Rivet Joint, from maintenance to communications to language to signals collection, analysis, and reporting. They were very clearly among the best the USAF had and knew it! It was "the dream team of the century."

But I saw the same thing I had seen with the EP-3E. Some of the senior NCOs at the bases we visited, generally the mainstay of any military organization the world over, were not particularly enthusiastic about this new "fancy" computer-driven system. They felt threatened. It was the younger, mid-grade enlisted guys that stepped up, just as it had been at Atsugi three years earlier. These young men had the advantage of three more years playing video games and using computer terminals. They were intrigued, not challenged, by the new technology and enjoyed both using it and figuring out what went wrong when it broke.

It was, yet again, an interesting leadership challenge. I did have to send several of the senior guys home as their attitude was just too negative, but most of the troops were there to make it work. At the end of the OpEval, I made sure they all got AF Commendation Medals, and for a few, the younger guys who had stepped up to the task, I was able to get Meritorious Service Medals. That the more youthful guys got the more senior medals would have never happened in the Navy, but this was obviously not the Navy, and I was learning how to play "Air Force."

RC-135W (with New Engines)

Another advantage was that E-Systems had already established procedures on capturing the needed systems engineering data when something went wrong. We reviewed them, liked them, and just kept the same processes in place. I did have a considerable advantage from my time on the EP-3E test team of understanding exactly why it was so important to capture all sorts of seemingly irrelevant data. When something went wrong, it was that data that often told us where the real problems were. That lesson has stood me in good stead for the rest of my career.

Our first operational flight was scheduled to be from Mildenhall, but many of the test team were not going to deploy forward; only about half were going. They were both the testing and training cadre. The other guys were disappointed they were not going to get the honor of flying on the first operational flight of an aircraft they had worked so hard at bringing into being. Then I had an idea. I discussed it with the Squadron Commanding Officer and Operations Officer. They were able to get permission to fly the first flight from Offutt as a mission against Cuba. It turned out to be a very productive mission. We even detected a new weapons system, previously unknown to be in Cuba, in an operational mode. We even sent a high-priority "Spot Report" alerting US intelligence to this new capability.

RC135W Rivet Joint Block III

After that mission, which was flown on the Wednesday before Thanksgiving of 1980, we all took a break before deploying to England a week later. On Thanksgiving morning, I got up very early at Offutt Air Force Base and caught an American Airlines flight from Omaha International to San Antonio. My wife and our two children picked me up at the airport, and we headed for my father's house in Corpus Christi, arriving in time for Thanksgiving dinner.

It was a great break, but what I remember most about that trip was sitting with my father watching the evening news on Friday and Saturday. One of the top news stories was that the Soviets had given Cuba a new, top-of-the-line weapons system. The US national leadership was upset about it. This news was delivered by an Assistant Secretary of State reading a statement to the press expressing the US's outrage. One of the announcement lines caught my father's attention. "In the past week, this new weapons system has been detected operating in Cuba."

My father's question was, "How in the Hell would we know it has become operational that quickly?" I just smiled and said nothing, but I absolutely knew the answer to his question. It was the weapons system we sent the Spot Report on two days earlier. The aircrew member who had detected it worked for me, and I had written the report! I can say this well into the 21st century but had I said this in the 20th Century, I could have gotten in a good bit of trouble. Times have changed.

On that trip, my father noticed my belt buckle with US Navy submarine dolphins on it. He was puzzled because I had never told him about my time in submarines. In those days, we just did not talk about that program. I made a bit of a joke about it and changed the subject. But the next time I was down, he told me he called my big brother, the commander in the Navy Supply Corps, and asked if he had any idea about what I was doing.

My brother had to report to my Father that he knew truly little about what I was doing. Still, he guessed it must be significant as it was highly classified. That was about all my Father ever knew about my career until Admiral Turner, then the recently retired Director of the Central Intelligence Agency, presided over my retirement ceremony eight years later. That, too, also completely blew him away. How Admiral Turner came to be the speaker at my retirement will be covered later in this book. It is an amusing story, at least it is to me. Bureaucracy reared its ugly head once again but got smacked this time. More later.

While we were at Greenville testing the Rivet Joint, the tragedy at Desert ONE happened. A USAF C-130 and a USN CH-53 helicopter, transporting part of the special forces team sent to rescue our hostages at the American Embassy in Tehran, ran into each other at the refueling rendezvous in the Iranian desert. All on board both aircraft died, and the mission was aborted.

Shortly after that, a number (8?) of specially configured C-130s landed at Greenville in the dead of night. They were not there when we secured work on the Block III late one evening but were there the following morning.

A big part of the vast (room enough for 17 Boeing 707s) hangers had been emptied, and the C-130s were immediately moved inside, away from the prying eyes of Russian and French satellites. However, in that our test airplane was on the apron right in front of the vast hangar and its door was open, we could look inside and see all these new arrivals. That morning word went out over the E-Systems private communications system. The newly arrived C-130 were not to be discussed outside of secure facilities. We were not to mention them to anyone, not even our families, nor were we to speculate about why they were there. Of course, speculation went wild. It was clear that we were making these airplanes into low-level flyers with terrain-following radar and adding defense systems such as flares and additional ECM gear.

Not long after that, a dozen or so of Army, Navy, and Air Force special forces (SOF) guys arrived at the Holiday Inn. This was the same hotel where my team and I had been living for the last several months. These guys were different from my intelligence systems operators. The SOF guys were all very fit and very tanned. We "spooks" thought we drank a lot, but the SOF guys quickly outpaced all of us "spooks." Not long after they arrived, one of our favorite waitresses came over to our team table at the hotel bar. She wanted to know just exactly who these new guys were. She could tell they were very different from us.

Our answer was, of course, the old classic "If we told you, we would have to kill you." The waitress was having none of that, and her reply was: "You always say that!"

"Yeah, but this time we mean it!" we replied. She took one look at us, took another look at the SOF guys, and decided maybe this time we really did mean it. Those guys looked and obviously were very tough. They were very fit, very tanned and all looked like they had some tough miles on them.

Shortly after that, I was read into their program. I was told that, because this would be a joint, all services event, I may well be asked to be in either the Command-and-Control aircraft or the Rivet Joint. Both planes were going to be orbiting over Tehran when the special forces teams went in to get the hostages.

This time we were going to go in extremely hard and try to destroy the entire Iranian Air Force on the ground at the very beginning of the operation. I have never heard anyone discuss the rest of that plan outside of the secure rooms at Greenville, and at CIA Headquarters, so I will not either, but suffice it to say the SOF guys were going to earn their pay that day. I did get a chance to talk with the SOF guys many times in the next few months about their mission and training and came away highly impressed. They are as bright as they are fit.

On Wednesday, 29 October 1980, six days before the presidential election of November 4, 1980, the mission was called off, and we all went home. I arrived in San Antonio on Friday at about 5 PM and tried to unwind by taking my two children trick-or-treating. It was Halloween, but nobody told the weatherman. It was sweltering, but not as hot as it would have been over Iran, where I had expected to be.

On Sunday, my wife and I went to church with our two children. The pre-sermon hymn that day was "Jesus shall reign where-ere the Sun." The second or third verse contains the phrase "the prisoner leaps to free his chains." With those words, my mind went back to the last two months of planning and training. I broke from the tension of those months. I started crying, thinking of the Americans still in prison in Tehran and the courageous men who were planning to jump into Tehran to save them. My wife thought I was having a heart attack as I fought to control my emotions. We got outside, and I calmed down, told her I was only tired from a lot of work, and we all went to eat Mexican food, as was our usual Sunday routine.

Two hours later, I was on an airliner headed to DC for meetings with the Air Staff and, on Tuesday, lunch with Admiral Turner at CIA HQ. On Monday, the Pentagon meetings went as planned, and I arrived back at my brother Bob's house in time for dinner with him and his wife. As I noted earlier, he was a commander in the Navy assigned to the Pentagon at that time. His main hobbies were vegetable gardening and cooking, using his vegetables. Meals were always great at their house!

As we began dinner, the phone rang, and Bob picked up the phone, and I heard him say "Yes, this is Commander Thomas" to a question at the other end of the line. Then,

"Yes, I will hold." Followed a few seconds later with:

"Excuse me, sir, but I am CDR Bob Thomas. You want to speak with my younger brother, LCDR Guy Thomas, who is right here," and handed me the phone with his hand over the mouthpiece and his eyes as wide open as I ever saw them.

"Guy, it is the Director, CIA, Admiral Turner himself!"

I took the phone to be told our lunch had been preempted by the British Ambassador and could I come half an hour earlier? Of course.

I turned around to one of the most quizzical looks I ever saw.

I had not given Bob any indication that I was visiting the CIA then, or ever, and I left him to wonder exactly what I was doing. In those pre-cellphone days, I had left his number as my point of contact, hence the direct call. Until that call, my brother had no idea with whom I was meeting or why, but you can bet that story went back to my father in Texas in a flash. (And my credibility in my family soared!)

Admiral Turner wanted my take on how I thought things would have gone down had we moved forward with the rescue attempt. I had participated in most of the planning meetings there at Greenville and several in the Pentagon. I believed I could say that the SOF personnel who would carry it out were sure they could do it. I thought it hinged on our ability to establish total air control over Tehran, especially over our egress route.

RC-135W, RIVET JOINT Block III

He thanked me and then said something very interesting, even exciting. This was election day 1980. He indicated that he would be moving back to the Pentagon in "a more senior role" if Jimmy Carter were reelected. He said he would have a place for me there after I finished my course at the Naval War College, for which I had just been selected if I was interested. Of course, I was! I thanked him and departed.

Jimmy Carter was not reelected, so none of that came to pass. A few years later, Geoffrey Turner, the Admiral's son, told me over dinner that Carter had promised his father Chairman, Joint Chiefs of Staff. I am not sure I would have made a good staffie, but the Office of the Chairman, Joint Chiefs of Staff would have been a fascinating place to work.

The week after Thanksgiving, we deployed the first automated Rivet Joint, the first RC-135W, also known as the "Block III" to RAF Mildenhall, where we started crew training for the entire squadron. Ten or so days later, we were ready to take Block III into an operational area to test its many new capabilities. We decided to fly in conjunction with the regularly scheduled RC-135V, a Rivet Joint Block I mission, to establish a real-time baseline.

It was my first ever mission over the Baltic, and it proved to be true to form with the rest of my "firsts." A day or two before we launched, Poland announced it was withdrawing from the Warsaw Pact, the Soviets counter to NATO. The Soviets were less than amused, and our pre-mission brief stressed that we were flying into an agitated situation. They also mentioned that besides the two RC-135s, the West Germans were also going to have one of their Atlantic reconnaissance aircraft in the same area. Interestingly, it was also built at Greenville, with much of the same equipment as a stock, non-automated RC-135V. I had even become friends with some of the German crew who had come to Greenville to pick one up and had gotten a tour of it.

Our advice from the intelligence briefer was that, if things went south and it did appear the Soviets were moving against the Poles, to abort the mission and come home. "No need to try to die heroes." Roger that!

We took off and trailed the regular RC-135V into the Baltic by about 30 minutes. When we got there, we also established secure voice comms with it and the West German Atlantic. As we turned north onto our mission track, we got a report over our secure data communications that the Soviets were loading parachute troops at airfields near Poland. The situation looked like it might go hot at any time. We acknowledged the message and then got two other reports over our voice system in quick succession. Both the other RC-135 and the Atlantic had been intercepted by Soviet fighters who were escorting them from astern, which was odd. Usually, they flew up alongside and looked over the cockpit crew.

SU-15 FLAGON

A few minutes later, our ELINT guys reported we had at least two Soviet Sukhoi-15 fighter radars locked on us. And judging by the signal strength of the radar signal, they were remarkably close. Even more concerning, the radar signals were

not being picked up by our sensitive collection antennas on either side of the aircraft but rather via the much less sensitive "gap-filler" antenna at the rear of the plane. This meant they were very close, within a few miles, and "in our 6", i. e, dead astern and in perfect firing position.

The pilot gently turned the aircraft first right, then back left. The cockpit crew could see two fighters 1,000 feet below us and 1,000 feet astern. He called me up to the flight deck to discuss what was going on. I had been doing the same with the mission crew. None of us had ever seen all airplanes in an area intercepted simultaneously, with no communications from the ground until contact was established. No comms meant we had had no warning they were coming. Also, none of us had ever seen the intercepting airplanes not close in for a visual examination of our planes. It was disconcerting to all of us. The pilot asked me what our warning of hostile intent and how much warning would we have. At this point, I had to answer that our first sign of hostile intent might be MISSILE IMPACT if they fire a heat-seeking missile, as we were now in the launch window of those systems. We would literally have no warning. They could launch on a codeword command, or maybe even at a pre-set time, such as 10:00 hours, which was in a few minutes. We were "meat on the table," as the saying goes.

We were already flying north, away from Poland. The pilot gently turned the aircraft further to the left, toward Sweden, and away from the Soviets. The fighters hung on for some minutes. I do not remember exactly how long it was, but it did seem like an exceedingly long time, which it was not. I do recall the relief we all felt when we detected them turning for home. The rest of that mission was pretty tame by comparison. All of us believed we had just faced certain death if Russia had decided to invade Poland. They would have shot us down and claimed they thought we were Polish aircraft or some such "hogwash!" All our families would ever get was an "Oops! Sorry!" Just like when the Israelis attacked the USS Liberty. It was a very tense 20 minutes or so, which I, and the rest of the crew, will never forget.

We flew the Block III for several more missions in the Baltic. Then we needed to fly a track around Norway's Nord Cape and into the Barents Sea/Archangel Operating area to test our systems in that unique signal environment. That was a quiet mission as it was already getting icy up there, but just before we landed, one of my best operators told me something that both froze and boiled my blood. He had detected some very odd communications mid-way down the Norwegian coast that he had not even bothered to record. He said it sounded like Soviet fighters taking off and landing on a ship, but we all know the Soviets do not have an aircraft carrier, right?

My reply, like my blood, was both icy and hot. I made the operator aware that it was apparent he was not reading current intelligence reports. The Soviets were preparing their first aircraft carrier for operations. The fact that it was as far south as he reported, well down the Norwegian coast, was hugely important as it showed they were further along than was thought. It might also mean they were going to try to sneak by England, much as we had done in the Sea of Japan with the HORNE task force in September 1975. It is also possible they might be going into the Baltic as a surprise show of strength to the Poles. Even though I had no proof and were well out of the operational area, I decided to send a Spot Report on this event.

The next day an SR-71, already tasked to fly a mission from England over the Barents, was charged to use its systems to try to detect the Soviet warship and did. The Soviet aircraft carrier reversed course shortly after that and returned to the Northern Fleet operating area, its home waters, without further incident. It is possible that it had detected the synthetic aperture radar of the SR-71 and knew it had been caught. What we had seen was probably a dress rehearsal for an "out of area" deployment.

What I saw was a significant need to tighten up the professionalism of my crew. We held an all-hands meeting, and I stressed that all of us needed to read everything we can about current ops, especially in the areas where we might be flying; we all had no idea when something which seems minor might just prove crucial. While I first learned that lesson on USS HORNE as a 3rd class Petty officer, it had been reinforced many times since.

We went home for Christmas. The Block III went back to Greenville to repair some airframe problems, tweak the mission system, and install some upgrades, especially to its air conditioning which had proven inadequate. Our computers gave off a lot more heat than initially anticipated.

The middle of February saw us ready to deploy to England a second time. We left right on schedule, and after less than three weeks, we determined that the repairs and upgrades worked very well. We were able to report the Block III ready for service. We had a little ceremony with me handing over a symbolic set of keys to the USAFSS squadron representatives, plus the leaders of both Squadron 38 (pilots and navigators) and Squadron 343 (Ravens- electronic warfare). Primary mission accomplished.

It was now fully ready, but I had a test report to finish. I had tasked the leading sergeant on each position to write "the good, the bad, and the ugly" for their particular system after each flight. I took all the inputs and edited them into what we all hoped would be a helpful document. The number one problem that many of us saw was that we now had the world's best electronic vacuum cleaner but no practical way to make sense of what was in the vacuum's bag and report it back out in a tactically useful time frame.

We all put our heads together to establish exactly what we wanted and how to get there. In the end, we came up with not only an in-depth identification of the problem but also a thorough description of how the problem could be rectified. This was my second pass at this problem, as we had seen a similar shortcoming with the first EP-3Es as well. We also listed what was good about each new position and sensor, as well as any problems with its interface into the overall system and what needed to be improved. We wanted to give the engineers the operator's view of what was right and wrong with the Block III.

Our two principal recommendations were related. First, there needed to be a sensor fusion system. Second, the fusion system product needed to be tied to a real-time downlink to both command/control systems on the ground, and to the AWACS airborne radar platform. The ground system needed to be linked to all available sensors and a wide-ranging database on all subjects pertaining to the area. This recommendation led directly to the creation of the Tactical Intelligence Broadcast System (TIBS), the forerunner of the Integrated Broadcast System (IBS), which I also worked on 15 years later.

General Larson asked me to develop a brief from my report, which I did. I gave my report briefly in San Antonio. Then he had me go on a speaking tour to brief the responsible folks at Greenville, Offutt, the Big Safari office at Wright-Patterson AFB, NSA at Ft Meade, and the Air Staff. I also asked for permission to brief the Navy on the test. After I finished with the USAF and NSA briefs, I briefed ONI, NavSecGru, and the Reconnaissance, Electronic Warfare, Special Operations, Navy (REWSON) office at NavalEx in DC.

I received a warm reception at the Air Force offices, but I could see that some of the Navy people were not impressed that a Navy guy (me) was so up on an Air Force system. They believed their design, the Deepwell system on the EP-3E, which they had upgraded and brought back into service, was better. It was not, not by a long shot. I know because I have been there.

Indeed, I tried for the rest of my career, and several years beyond, to get the Navy to buy the RC-135V/W. Failing that, they should at least buy the same system and stuff it into the P-3 airframe. The important thing was to use the same consoles and software on all platforms for standardization between services with the same mission. What a concept! But it was not to be in my 30+ more years of professional life.

I did have two events that stick with me to this day, nearly 40 years later. One was my reception by the Air Staff. They were highly complementary and very gracious. I forget whether Jim Clapper was still there, but I think so. I know he did "make his star" and move on about that time. They all saw what I had done as something of national importance and said so.

However, it is another event that sticks with me to this day. It happened at SAC HQ in Offutt. I had just finished my brief to CINC SAC, the four-star general himself, and his staff, including a 3-star and four or five 2- and 1-star generals. One of the colonels I had been working with for most of the last three years asked if I could join him and a few others in the cafeteria for a cup of coffee. I had a full schedule that day, but he seemed to think it was necessary, so I agreed.

We walked down to the main cafeteria, which was nearly full. However, the Colonel guided me over to a table with four more full bird colonels I knew. I had been working closely with their organizations since arriving at Offutt over two years before. As I approached, they rose as one and, donning their hats, formally saluted me. It had obviously been planned and rehearsed. It was a huge compliment, unique in my experience. I had never even witnessed anything like that, much less been the subject of it. To this day, thinking of it still chokes me up a bit.

Their words were less grand. Basically, they were to the effect that I had not done too bad for a Navy guy. And thanked me for showing them that their idea that the only competence in the US Armed Forces resided in the USAF was not altogether correct. Of course, that is the Air Force way of saying "Well done!" the highest praise given in the Royal Navy.

We all congratulated each other for bringing Block III online. In truth, I could not have done it without their help and that of a lot of other folks at each of the places I visited on my exit briefing tour. We all laughed at each other and shook hands, and I left to brief the 55th Strategic Reconnaissance Wing, then fly on to Wright-Patterson AFB, then to Ft Meade and DC to give more briefs. It was a fantastic way to end my tour with the USAF.

One interesting note was my reception at each of the hotels at which I stayed on that tour. The folks at each of the reception desks all said much the same thing. "Welcome back! You spend more time here than you do at home!" And they were all correct! I had spent 27 of the last 30 months away from home.

My time with the AF was ending. Not long after I returned from giving the status report/out brief on the Block III, the USAFSS (recently renamed the Electronic Security Command (ESC) HQ had an awards ceremony. It was a formal affair on the parade ground, and all hands were encouraged to attend. At the end of the ceremony, the time slot traditionally reserved for the most significant award, MGEN Larson pinned USAF wings on my US Navy uniform.

NAVY OFFICER EARNS AF WINGS -- Maj. Gen. Doyle E. Larson, ESC commander, pins the Air Force Officer Aircrew Member Badge on Navy Lt. Cmdr. G. Guy Thomas, chief, Line of Sight Operations, HQ ESC. Commander Thomas qualified for award of the USAF aircrew wings by flying more than 400 hours aboard Air Force RC-135 aircraft. As test director in the development of the Rivet Joint Block III, the newest in RC-135 modifications, Commander Thomas is the first member of the Department of the Navy to be awarded the Air Force wings.

I subsequently discovered that this was the first time the US Air Force had given a member of the US Navy, Army, or Marines, Air Force wings. After an interesting exchange of letters with the Navy Uniform Board and then a visit to their office in

DC, because a US general officer had pinned them on my Navy uniform, I was allowed to wear the wings under my ribbons while I wore my Navy wings above them.

Over my last seven years in the Navy, it was pointed out to me many times, and in many places, that I was out of uniform as "it was not permitted to wear insignia earned while in other services on the Navy uniform." That is true, but these wings were pinned on my <u>Navy</u> uniform by an Air Force general; thus, I was permitted to wear them. That made them a unique honor.

To be honest, I had hoped for a Legion of Merit (LOM), and all three of the colonels I worked with told me they had argued for me to receive one. A fourth colonel I did not even know, the head of the ESC's awards board, looked me up to tell me he had forwarded the award recommendation for a Meritorious Service Medal for my time as Director of the Block III acceptance test to the general saying the award recommendation was the strongest, he had ever seen for an O-4 working not one, but two pay grades above his rank and recommended it be upgraded to the LOM.

The fact that I was an O-4 (lieutenant commander) serving with significant success in a senior O-6 (colonel) position of high national importance was highlighted, as it had been in my last two annual Fitness Reports, and that had made a distinct impression on him. He did not realize that both the Director of Operations and the Director of Acquisition had separately and independently recommended to the General that I be given a LOM as well. In all three cases, the General had replied that I had been given AF wings, and that award was enough. Oh, well. Such is life. The Head of Operations, for whom I worked directly in my role as the Chief, Line of Sight Collection, decided he would also put me in for a second MSM as an end of tour award to make up for the fact that I did not get the LOM, which did assuage my disappointment slightly.

About four-fifths of the way through my time as test director, the Current Operations Technology Advisor, a GS-15 civil servant, retired. It was the #3 position in authority in the Directorate. They wrote a job description and gave me a copy. "You know anybody that meets this description, please let us know." I read the document and instantly realized I had all the qualifications they were looking for but decided I was having too much fun in uniform to give it up. Besides, I believed I was still in the running to make Admiral. Especially since I had been selected to attend the Naval War College, a dream of mine for years, and had competed with 111 others, three year-groups of NavSecGru LCDRs, to win that selection.

They selected someone else out of NSA for the Technical Advisor job, and while he was a nice guy, he had a steep learning curve to get up to speed. I am not sure he ever made it. Several months later, Colonel Jerry Wuscher privately expressed his disappointment over the selection. I remarked that I had considered applying for that job. His reaction startled me. He told me that the requirements were written with me in mind, and the front office had hoped I would "rise to the bait." I should have taken that bait!

They had plans for me to continue as a Naval Reserve officer and as a GS-15. The AF did not want to actively recruit me as that was well beyond the bounds of propriety, but he told me they were disappointed I had not applied. Had I become a GS-15 at 36 in that command, I suspect I would have been a subsequent strong contender for Senior Executive Service. If I stayed in the reserves, I would have almost certainly made captain and been a solid contender for admiral as well, given the close relation my civilian job had with my specialty in the Navy. And I love San Antonio!

It would be another 20+ years before I became a GS-15, and for sure, I never even went before the selection board for captain, much less rear admiral. I could have retired at age 62, making, in retirement, about what I was making working full time at that point. Oh well, hindsight is hindsight, and I did like my last eighteen years of my professional life, nine years at JHU/APL, and nine years with the US Coast Guard in the Maritime Domain Awareness Program Integration Office (MDA/PIO), an office with a national mandate. Indeed, I was enticed to that job by being told we worked directly for the National Security Council. As I have said elsewhere in the memoir, I know I have been incredibly blessed for some reason.

I transferred to the Naval War College in July 1981. I was very much looking forward to it. It had been a dream of mine since before I had joined the Navy!

Naval War College & its Center for Naval Warfare Studies

Dreams Really Do Come True!

Chapter Eleven

My family and I arrived at the Naval War College at the very end of July 1981. All of us were excited to be there. Newport is a great place, especially in the summer. It was 80 the day we arrived from San Antonio, and everyone was complaining about the heat. It seemed like a lovely Spring day to us, coming from Texas!

Naval War College, Newport, RI

We quickly moved into the on-base housing reserved for the Naval War College, and I got right to work as a student. I was assigned to a seminar with a professional educator as its lead. One of the first things I did was look over the complete curriculum for the year. It is broken into three trimesters: Strategy & Policy, Naval Operations, and Quantitative Analysis.

I looked at the curriculum and noted that nowhere was there any mention of either Space or Special Operations Forces (SOF).

From my time with the Air Force, it seemed obvious that both Space and SOF would be especially important in the coming years. I asked my seminar lead if he had any idea why they were omitted.

He guessed that they were both too highly classified to be taught in any meaningful way at the NWC, which conducted its classes at the SECRET level at most, with much of the curriculum being taught at the UNCLASSIFIED/For Official Use Only level. I begged to differ with him. I agreed that day-to-day operations and specific capabilities were, indeed, highly classified. Still, the concepts and fundamental organization of both Space and SOF were unclassified. And a great deal could be included in a course at the SECRET level. I further made the point that the NWC was supposed to be preparing its students for future conflicts. Both Space and SOF were growing in importance. The NWC should at least include introductory sessions on both.

My seminar lead bought my reasoning. He made an appointment for us to discuss this with the Director of Curriculum, CAPT Sullivan. He had a P-3 background, and I was wearing wings, so we had something in common but I had almost the exact discussion I had had with my seminar lead.

"Too Classified"

"Not the concepts and you are teaching concepts here in most cases. And they are both going to be more important in the future."

"I take your point. I will investigate it. Thank you for your suggestion."

I went home feeling I had done my good deed for the day, and maybe next year or the year after that, they would add Space and/or SOF to the subjects taught at the NWC. Thus, I was a bit surprised about three weeks later when I got a note from CAPT Sullivan asking me to come to see him at my earliest opportunity. I made a point of going to see him as soon as I could. It was a fascinating conversation.

"Checked with both the SEALS and Navy Space. Both organizations think it is a great idea and are a bit sheepish that they had not thought of it themselves. However, both commands are already overcommitted and cannot spare anyone for two years. They both have suggested we pull one of our students out of classes to be their stand-in until they can get someone here. Interestingly enough, it is the same person."

My mind must not have been fully engaged at that moment as my curiosity got the better of me, and I replied: "I should know that person as I worked with both communities in my immediate past tour. Who is he? I bet I know him."

His reply was classic Naval aviation. "I bet you do! Your wife is sleeping with him!"

"Huh? Who is this SOB?" Then the light came on, finally. He meant ME!

He did not want me to withdraw from the student body as it would mean I would not graduate, which would be detrimental to my career, but he did ask me to start setting up two study plans for multi-day sessions on both Space and SOF, and I did. To this day, I am immensely proud that I brought both Space and SOF to the Naval War College.

As a history major in my undergraduate studies, I most enjoyed the Strategy & Policy semester at the NWC. Starting in elementary school, I had done a great reading on history and politics, and the course at the NWC put a structure to it. We had a massive amount of assigned reading, and I tried to get through it all, but I suspect no one did, or even could. I did OK on my papers, but I was most pleased with my "end-of-course" verbal grade and the attached remarks, highlighting my knowledge of the course material. They also noted I was willing to listen to other points of view. My instructor, a professor of some renown, stated he had come to consider me as a colleague rather than a student under instruction. High praise, indeed.

During that trimester, I also monitored the Introduction to Intelligence elective. With my background, they would not let me officially enroll. Still, I knew there was much about intelligence other than SIGINT that I did not know, so I insisted and was permitted to monitor the course. However, the course was taught by a human intelligence specialist with some experience in airborne imagery and little else. After a few sessions, I offered to conduct a class in tactical SIGINT, which he let me do. I also noted that there was no section on foreign intelligence and electronic warfare threats. These were favorite subjects of MGen Larson, my immediate past boss, and I wanted to look at them from a Navy viewpoint, so I asked if I could do my term paper on them. My idea was that my study would become an outline for a future class. His answer and subsequent actions completely changed the rest of my time in the Navy, but I am getting ahead of myself. It was yet another unintended consequence.

He agreed, and I started researching the many documents held in the classified library at the War College. There was a fair bit there, and I became good friends with the library research staff. They liked the idea that I was looking at books and journals that, according to them, had rarely been accessed. In the end, I came up with four terms new to me, plus the information on Soviet *radioelektronnayabor'b*a Radio Electronic "struggle" or "combat" in English. (REC), the principal subject I was looking for. The new terms were:

OsNaz, otryad osobogo naznacheniya: roughly the Naval Security Group of the Soviet Navy but may mean something else in the other branches of the Russian armed forces and security elements.

Spetznaz, spetsiálnovo naznachéniya: the special forces of Russia, there were many more of them than there were in the US military.

PKB, Protiv Kosmotichkii Borba: "Anti- Space Combat"

with its own *PKO, Protiv Kosmotichkii Oborny* "Anti-Space Organization," a fully developed concept including its own doctrine and full-up organization in Russia.

Of particular interest to me was the fact that the PKO had a section on REC. I immediately realized that the Russians were organized to conduct electronic warfare against our space assets. I knew of the REC concept in general terms. It was what I had started looking for in the first place, information on Soviet electronic warfare capabilities, both active and passive. Indeed, the Soviets had a fully developed doctrine, broken down into Strategic, Operational and Tactical components. But what was new to me was that they were planning to attack our satellites via electronic warfare. This was news to me and nearly everyone I talked to in the next four years. More on this later.

I enjoyed my research in that little-used archive and, in the end, had my paper. I was enormously proud of it. To the best of my knowledge and those of the library research staff, these were all new subjects at the NWC.

I submitted my paper on time and eagerly awaited the instructor's comments. However, I was shocked when the document was returned to me. It had a rude note scribbled across the title page. It read in effect that I had obviously done a great deal of work on the paper, but it was too technical and full of "technical jargon" and unproven (sic) concepts that the piece was impossible to read and was useless. He declined even to grade it!

I was utterly crushed! As a history major, I had written a good deal in college and then in the Navy and believe my command of English was well above average Indeed, I had been often told my post-mission reports from both submarine and aviation missions, in both the USN and the USAF, were very clear and easy to read. So, I was not prepared for this total rejection of what I believed was the most meaningful report I had ever written, with the possible exception of the Operational Test and Evaluation Report on Rivet Joint Block III. That report resulted in the creation of the Tactical Intelligence Broadcast System (TIBS). But at that moment, I believed my just rejected paper was even more critical.

I went back to my desk in a very melancholy mood. I was considering what to do next when the Navy public affairs officer who had the cubicle next to mine walked by. He saw I was in some distress and asked what was wrong, someone died? I replied by throwing my paper to him and saying something to the effect that he should read the note on the front page. And I commented that "If this is too technical a paper, then the Navy is in trouble!" He scanned it, said it looked interesting and asked if he could take it to read in detail? "Be my guest. I have no further use for it! I never want to see that paper again."

A short time later, he returned to me and cheered me up a great deal when he agreed it was not too technical. My "jargon" was common modern operational Navy lingo. Indeed, he definitely liked the paper and thought a fair bit of it was unique to his experience.

In that he had already taken the Strategy & Policy trimester via correspondence, he was working for the research side of the NWC, the Center for Naval Warfare Studies (CNWS). He told me this was just the sort of thing they were looking at there, and would I mind if he took my paper over to them? I, substantially heartened but still skeptical anything would come of this, agreed.

A day or two later, the director of the Center for Advanced Research (CAR) at CNWS, Lieutenant Colonel (LtCol) Orville "Bud" Hay, USMC, sent me a note, asking me to come to see him. Bud was a Marine fighter aviator and the most influential 0-5 I ever met. (Think of an even more articulate Oliver North with a highly analytical mind.) I subsequently learned he was known and highly respected at the Navy and Marines' most senior levels.

He had read my paper and liked it very much. He asked me to drop out of the student body for the next semester and become a Research Associate at CAR. My work there would count toward graduation. The next semester was "Naval Operations," where they examined how the Navy's air, surface, and submarine branches worked separately and together. They also studied how requirements became acquisition programs. I now had significant experience with all these subjects. So, knowing I would not be missing much, I readily agreed.

Additionally, Bud Hay was assembling a team to look at the capabilities, shortfalls, and vulnerabilities of the Navy's communication and intelligence collection systems. In that a good deal of this involved a close interface with the USAF, he thought I would be the perfect person to lead it. He also wanted the team to make recommendations about how this could be input into the nationally significant war games. These games were conducted many times a year at the NWC's Center for War Gaming, another part of CNWS, as a means to test the concepts being developed by the Strategic Studies Group (SSG).

I was subsequently joined by an NSA civilian, an ONI civilian, Dennis Callan, and Major Jack Camm, a USAF technical intelligence officer. All were very bright, and we got right into our task. We worked well together and, by the end of the trimester, had our report ready. It was primarily a SECRET document, but there was a TOP SECRET Annex. We were proud of our writing but knew more needed to be done and said so in the report. Still, for a 3-month research effort, we had uncovered a substantial number of incredibly significant vulnerabilities in our C3I systems.

During the time that we were working on our report, SSG-I was halfway through its year-long look at where the Navy was going, and we interfaced with them every day. We had many lengthy discussions with the entire group, for whom we were basically working as research assistants. Our most in-depth and searching discussions were with Commanders (CDR) Arthur "Art" Cebrowski and William "Bill" Owens. All of SSG 1 were standouts, but these two men were both the most junior assigned and probably the brightest. I distinctly remember thinking that all eight members would probably make a flag, but both Bill and Art would wear three stars at a minimum.

Both indeed made vice admiral, and Bill, who served as the Executive Officer (XO), the 2nd in command, on USS POGY right after I left, went on to wear the four stars of Admiral. He retired as Deputy Chairman, Joint Chiefs of Staff. Art developed cancer and died at the height of his career. But his impact is still felt today, and the Information Technology Research Center at the Naval Postgraduate School is named after him for his work in developing the Network Centric Warfare concept. My team was in good company. Had he not died, it is clear he, too, would have made four stars.

From those discussions, I developed my concept of "Warfare in the Fourth Dimension," an idea that has caught on all over the world. However, it was others who took the model and really pushed it forward that made it happen. I did publish three articles on the Naval War College Review based on my research of late 1981 and early 1982. The first was on the threat posed by Soviet Radio Electronic Combat (REC), where I outlined the threat it posed to the Navy. The second was on "Warfare in the Fourth Dimension," which outlined the need to control the electronic spectrum, equating it to the need to control a battle timeline. Time is the 4th dimension of physics, and the 4th dimension of warfare is electronic warfare.

The newly formed Strategic Studies Group (SSG-I) was eight officers, six 0-6s (five Navy captains, one from each major warfare specialty, and a marine colonel) and two commanders, Art Cebrowski, a fighter pilot, and Bill Owens, the only submariner.

Art, Bill, and I became good friends. They were both very interested in the many aspects of C4ISR (command, control, communications, intelligence, surveillance, and reconnaissance), and that was where I had just spent the last 11 years of my life.

We had many discussions on both our C4ISR capabilities

and those of our opponents. These talks dovetailed exactly into my previous trimester research, and Art and Bill were highly interested in it. The other members of SSG-1 and those of the next several were also interested in C4ISR, so I became their de facto subject matter expert (SME) and stayed in that role for the next several years.

It was an exciting time as the SSG was developing "The Maritime Strategy." This, too, was a watershed point in my life. Subsequently, I gave up an opportunity for command, for which I was selected. I had aspired to command for most of my life, but I elected to remain at CNWS and continue development of the Maritime Strategy. It was that interesting! CAPT Sam Leeds, one of the brightest lights of SSG-1, did me the honor of speaking at my retirement from the government in 2012.

I was one of the few NavSecGru officers who published anything in either the NWC Review or the Naval Institute Proceedings. We just did not usually call attention to ourselves in those days. We lived "In the Shadows!"

My team and I went back to the "Quantitative Methods" trimester. Shortly after, we got our grades for our research work, which counted toward our class standing. We were both disappointed and outraged. My grade was 93, five points less than my S&P grade! The others had similar gaps, but mine was the most significant disparity, probably because I had a 98 (out of a possible 100), one of the class's highest grades. No one made 100, and only a very few made 99 or 98. I believe every other team member had at least a 95, and they all received 91s or 92s. All of us had worked very hard on something we thought was very significant. We knew we had produced an excellent, useful product, and this was our reward? To be knocked out of contention for Distinguished Graduate?!?

I immediately went to see Bud Hay and asked what was going on. He pulls four of the brightest guys from the class. He works them like dogs. They really produce for him. Then he gives them significantly lower grades than they had in the previous semester, thereby ensuring that none of them will graduate in the top 10% of the class and make "Distinguished Graduate"? Say, what? We should each be given at least the marks we had made in the previous semester, if not a point or two higher. That seemed more than fair to me. Indeed, that was what my team and I thought would happen.

He understood our concern, but he was caught in a trap. Last year he had given the research assistants high marks, and the factuality had complained, so he tried to rectify it with our grades. I felt we had been cheated and told him so, in no uncertain terms. He pointed out that we had identified several areas for further research, so the report was not completely finished. I pointed out we had only about two and a half months to do our research and two weeks to finalize the work. I believed we had done an excellent job in the time given us.

He bought my reasoning and reviewed the grades, moving them all up a few points. He also asked me, alone, to spend my elective time at CAR, working on expanding and polishing our report, which I did. It was also about that time that it was decided that I would be assigned to the Center for War Gaming as the signal warfare assistant when I graduated from the school. This was another true honor.

The "Quantitative Analysis" trimester was the analytical portion of a year of the Harvard Business School MBA program compressed into three and a half months. With the work at CAR and the regular curriculum's workload, the time flew by very quickly. I did graduate with Distinction, which I am immensely proud of, considering the competition. There were only two naval officers in my class with a higher grade point.

Bud's boss, the Honorable Robert Murray, the Director of CNWS, invited me to his office the day before graduation. He let me know that our report had finished as a very close second to a research paper written by a single individual for the Mittendorf

Prize. That was the award for the best research paper that was written that year at the NWC. Close, but no cigar! I still cherish my time as a student at the NWC as one of the most rewarding periods of my life. However, I soon discovered that things were only going to get even better there at the NWC.

In July 1982, I moved down the hill to Sims Hall, a 19th century Navy schoolhouse and barracks converted into the world-famous Center of War Gaming (CWG).

Established in 1886, the CWG it was the first naval wargaming center in the world. It was housed in Sims Hall, a converted 19th century Navy schoolhouse and barracks It was where "the rubber met the road." Where the ideas generated throughout the CNWS were tested and examined for relevance and feasibility. What a great place to work!

The CWG had an intelligence detachment whose primary task was playing the opposing force as accurately as possible. They were the "Red Team." They were cautioned against playing the opponents as giants or as midgets. After a short time as a wargamer, I was formally assigned to the Red Team as their Space, Intel systems, EW and SOF SME. As a CWG staff member, I was trained on developing and executing a wargame and then on directing a war game I was also be responsible for the accurate portrayal of both Space systems' and SOF capabilities on both sides, ours and theirs. I also had responsibility for the correct portrayal of foreign electronic warfare and intelligence collection systems. I recruited members of both the Navy Space and SEAL communities to help me as needed for specific war games. However, my portrayal of Soviet, Chinese, Korean, Cuban, Libyan, and all other Soviet client states' capabilities were based on my current and ongoing research.

I was given access to all the US intelligence agencies people researching my areas of responsibility. To a person, they were very willing to work with me as I was a primary conduit for their research to reach the many decision-makers who regularly attended these war games.

In those hectic days, CWG held nearly a war game a week. The crowning game of the year was the Global War Game, Bud Hay's creation. It took place in the three weeks between the departure of one class and the next class's arrival. That was in July, a beautiful time to be in Newport. All services, all intelligence agencies, and most departments (State, Defense, Commerce, Justice, Transportation, Energy) attended.

The focus was how do we wage a protracted, non-nuclear war with Russia and the Warsaw Pact. How do we keep it from going nuclear and still prevail was a significant focus.

In the first game I attended, July 1982, Space and SOF were discussed. However, there were no tools to help understand or show how the capabilities and impact space systems and SOF, or lack thereof, would make in this scenario. Both the Army and Navy's TENCAP offices attended. Maybe the Air Force's TENCAP did, too, but I do not remember them at Newport for another couple of years. I do not recall any SEALs or other SOF forces there either, but both space cadre and SOF did come to play with a vengeance later in my tenure, and I was credited for making that happen.

In July 1983, the National Reconnaissance Office (NRO), through Navy TENCAP, offered to build a space systems simulator for CWG. I became the NWC project officer, working with Navy TENCAP over the next several months to get the requirements right. The money was coming via an NRO element which an Army officer managed. He and several people from the company selected to build the system visited the NWC several times. Every time they did come, their visit caused a stir. They needed to understand our physical layout and installation plans.

However, none of the men they needed to meet had more than a secret clearance, and they wanted to hold these meetings at three levels beyond Top Secret.

Common sense won out, but it was clear the NRO members were nervous about putting a space systems simulator in a facility that planned to port out the data to a secret-only facility. The data processor was in an appropriately secured and certified room, but the idea of downgrading its output to the level where it could be used by most of the operational users was just foreign to them. Indeed, this exactly reflected the real world at that time, and our lessons learned while dealing with this problem were genuinely ground-breaking.

We also had a serious discussion about how to build a system to meet a future projected capability, as most of our war games are set in the future. The NRO had highly detailed models of current US systems, but we believed we needed something much less complicated, especially to depict potential adversaries' systems. It was an intense set of discussions, but we eventually won. The simulation system, as delivered, allowed us to define its orbit, the swath of its sensor, and the time from tasking to collection to the information from the satellite arriving at a decision maker's desk. When the US's Space Warfare Center was created at Colorado Springs five years later, they selected a slightly updated version of what we created at the CWG as their core simulation system. This is something else I am proud of to this day.

When we accepted the system, we published a secret message to the US military at large describing its capabilities. Later that day, I received a call from a Navy commander at the NRO. He was irate! According to him, I had committed a significant security bust. In listing the types of satellites we could model and insert into a game, I had included radar satellites, thinking of the Soviet's Radar Ocean Reconnaissance Satellite. Our understanding of its capabilities was classified SECRET in those days. Most of what we knew about it at that time was subsequently downgraded to UNCLASSIFIED, at least in part because of the experience gained from those games. However, the angry commander said it was against all security rules to discuss radar satellites.

At that moment, I happened to have a copy of the then-recent Aviation Week & Space Technology issue, which described the Soviet radar satellite system in detail, on my desk in front of me. So, my reply was straightforward. If I had created a security bust, then AWST was indeed also guilty, right?

He asked what in the world I was talking about. I gave him the title of the article, its page, and the date of that issue. I also pointed out that that was the system to which I was referring. Did he have another system in mind? He mumbled something about how he would investigate it and hung up. Obviously, he was thinking of a program I was not even aware of at that time, much less cleared for. That was the last I heard of that issue.

Until 2018 or so, I would run into people with satellite-related clearances who are utterly ignorant of how far foreign and especially commercial space systems have come. They would get very uneasy when I start talking of these unclassified commercials and foreign governmental systems' capabilities.

I also continued to lecture on Space systems at the NWC, often in conjunction with an actual member of the Navy Space "fraternity." My prime draw was RADM Denny Culbertson, the Director of Navy Space. He and his wife loved Newport, and he came up every chance he could. During this time, the Navy decided to establish a space cadre by creating a "designated subspecialist in Space Operations." The Navy asked all officers who met specific qualifications and wished to submit a request to receive this warfare designation. The list of those selected for this new designation was published shortly before one of RADM Culbertson's visits to Newport. Upon arriving in Newport, he asked me why my name was not on that list. I had to tell him I had not even thought to apply as I had never served in an operational USN billet dealing with space as specified in the call for nominations. All my official "space-time" was USAF!

He said I knew as much about Space as nearly anyone else in

the Navy. (I had now been lecturing on Space systems and inserting them into war games for almost two years at that point.) He also said that the Navy needed me as a space professional. He gave me a direct order to apply immediately, that night, as soon as possible. He would speak to my boss to be sure it was expedited. My only option was to say, "Yes, Sir. I will get right on it!" And I did. I sent off a letter request with a nice endorsement from my boss at the NWC a few days later.

About ten days later, a supplement to the earlier announced list of new space subspecialists was published. There was only one new name on that list, mine. So that is how I became a Space guy. That also made me the first person in the Navy to have earned all four designations. Surface, Air, Sub, and Space, but it was via about as unique a path to get there as I ever heard of.

I also hired a very bright naval officer with significant modeling experience to build a model of all the SOF operation elements and assign a variable probability of success based on the inputs. We used that tool as a guide, but it did seem to be too harsh as most missions failed for one reason or another. However, it did have the value of acquainting everyone who used the model as to how complex a SOF mission was. Soon, SEALs and Army SOF were frequent participants in our games. They mostly came in planning roles and as subject matter experts, so we created the venue and method for SOF to interface with many other armed forces' elements systematically. The Army Brigadier General (BGEN), who taught SOF at the National Defense University, subsequently took our tool and spent far more than we did, making it much more robust. It was in use at NDU for a number of years. It may still be there for all I know. He did give us full credit for starting the tool. I was proud of that, too.

Global War Game V (GWG V), July 1985, was a turning point in both Space and SOF, both of which were played strenuously by experts in those fields. Game highlights in the last part of Week 1 through Week 2 focused on the damage that was being done on both sides by small SOF teams. The attention given them was all out of proportion to the damage they were inflicting. SOF and their capabilities and limitations were now a significant discussion topic in most, if not all, war games at Newport. Mission accomplished.

Also included in every day's brief at GWG V was the drumbeat of reports on how the Soviets were working very hard trying to take out our space reconnaissance capabilities via direct ascent missiles, lasers to blind them, and various forms of electronic warfare. We also injected as to how the Soviets were trying to destroy our ability to use satellite communications systems and were trying to cripple our internal transportation system via electronic warfare. This was in the mid-1980s. We called it "deception" then, but today it is called hacking. We judged that the Soviet effort had been largely successful against our space reconnaissance systems.

On Wednesday or Thursday of the 2nd week (GWGs are three full weeks long), my deputy in the Space Cell, an Air Force Colonel with substantial space experience, stood up during the morning intelligence brief and asked the briefer where he had gotten the information he was briefing as all systems that could have collected that information were now disabled, if not destroyed. The reaction was immediate by all the senior officers there, including the Chief of Naval Operations. "This cannot be!" and "It is impossible. Space is invulnerable." were the typical remarks.

About a week after the GWG was over, a Captain on the CNO's Executive Panel, I think he was a surface officer, came to see me from DC. The idea that we might lose space systems had gotten the CNO's attention, and he was here to check out why we believed as we did. We had a frank and open discussion, and I think he went away worried. Several days later, my boss informed me that the CNO had convened a meeting in Newport for next March to discuss space systems vulnerability. The three Rear Admirals most involved in space either as providers or users, and their staffs would be the principal attendants. My boss and I were to be their host.

We set up the agenda as requested by the three Admirals and prepared for a packed house at the meeting in our classified meeting room. I arrived over an hour before the meeting was to start just to make sure all was ready. RADM Tom Betterton, a legend in Navy Space as the father of one of our more effective space systems, was already there. I had never met the Admiral before and indeed did not know who he was at that moment. Still, since we were the only two people there, I went over to him and introduced myself as one of the hosts. He introduced himself, and of course, I immediately knew who he was, and I was glad to finally meet the famous man. However, the next two minutes were among the most trying in my life. His first words after the introductions were to the effect that he knew who I was. I was "THE jerk" who had made his life miserable for the past several months. He went on to say that my analysis was very flawed, and his systems were nowhere near as vulnerable as I had led people to believe.

I could have stood there and taken it, but at that point, I had been looking at this problem for nearly four years. I had read everything I could find at all known security levels. Also, I had interviewed every person with a Space background who came to the NWC as a student or as a participant in a war game or to speak for all those years. I had also interviewed the head of the NRO and several staff members, none of which were Navy.

In those years, I had also talked to the space systems analysts at both ONI and Air Force Intelligence at length. If I did not know what was going on regarding the vulnerabilities of our space systems, it must be a very dark secret, as I had all the proper clearances, or so I thought and told the Admiral so. He harrumphed and said something to the effect that "We shall see." ***It was the low point of my career!***

The meeting lasted for almost two days. There was a lot of discussion as to vulnerabilities on both sides and the need for more analysis. I believed I had accomplished my mission of raising awareness of the importance of Space and the Navy's need for it. There was also the significant added benefit of increasing the recognition that some of these systems may well have substantial vulnerabilities.

One of the things I liked about the Naval War College was the opportunity to meet remarkable people. This was true as a student, but even more so as a staff member. The three most interesting were famous in their sphere, but probably the most famous was Tom Clancy. He and I were at a dinner for authors at the Officer's Club there at Newport. He had just published "The Hunt for Red October," but he was still unknown and humble at the time. He was standing in a corner looking lost. I had no idea who this forlorn-looking fellow was. But, as a host command member, I thought I should go over to him and make him feel welcome. Glad I did!

At least in part, we hit it off because he asked intelligent questions and kept asking for more information until he fully understood the answer, just like I have always tried to do. He was fascinated that I worked in the Center for War Gaming and asked many questions about it. I particularly remember two questions. One was why the Navy was buying Harpoon missiles with a 60-mile range for its submarines when they had no way to target them. I fully understood how they were targeted. We had played that capability in unclassified war games, so I explained how they were targeted using a combination of acoustic and passive electronic warfare systems. He subsequently used exactly the methodology in "Red Storm Rising."

He also asked about the range of Soviet fighters and whether they could reach Iceland. I pointed out that the "range" you generally heard quoted was combat range, which is a radius, not a straight line, with time for an engagement in the middle. Generally, the ferry range was more than twice the combat range. "AHA!" Was his response. That too figured in "Red Storm Rising.

He asked to see the Center for War Gaming (CWG) and the automated wargaming floor, which was unclassified when a game was not underway. I knew we would have a demo for this group, but I could see he wanted a more in-depth, detailed tour. We made an appointment for the following morning but just before that time, the Dean of Research, the Director of the Center for Naval Warfare Studies (CNWS), asked me to see him in his office on the far side of the campus. That is not a standard request. He was the second senior person at the NWC behind the Admiral. A request to come to see him as soon as possible was not something you could refuse, especially if you were not sure you had a firm appointment with an unknown author. So, I asked one of my colleagues to show him the game floor if he did show up. My colleague did not understand the importance of what we did there and gave him a half-hearted brief, having no idea who he was. My main regret in missing him was that I missed my chance to explain the importance and impact of the CWG on Navy Planning. Indeed, the CWG impacts many different parts of the USA's government.

Tom Clancy figured in my life one more time. His fourth book was "The Cardinal of the Kremlin." Major themes in the book were the Soviets' counter-space capabilities and their special forces, the Spetsnaz. As noted above, these were the two main areas of my responsibility for insertion into war games. I led the research in these areas for the school. There were capabilities portrayed in Tom Clancy's book that I had never seen discussed outside of classified vaults inside of classified vaults (sic). They were not just Top Secret. They were compartmented. For several months I slept very uneasily. I fully expected to have both the FBI and NCIS come knocking on my door any day to ask me if Clancy and I had talked about any of this, which we had not. That is my story, and I am sticking to it! Sometime later, I was told the Secretary of the Navy had those facts unclassified to allow for counters to be funded and developed....

I also met Norman Polmar, the famous authority on both the

Soviet and US Navies. He has written many well-known books on both. I was invited to lunch with him by Frank Uhlig, the Naval War College Review editor and publisher of the War College Press. I had written a couple of articles for Frank, and he thought Norman would like to meet me.

There were just the three of us. Frank and Norman were old friends, and they had a lot of "catching up" to do, so I was mostly just listening to these two very erudite men until Norman mentioned my article on Soviet "Radio Electronic Combat," complementing Frank on publishing the most unique piece he had seen on the Soviet Navy in years. Frank acknowledged the compliment, saying he knew Norman would like it and would probably like to meet the author. Norman replied: "Absolutely," and Frank pointed to me, saying: "There he is!"

I was suddenly transported from the "wings to center stage," as a performer would put it. Norman wanted to know where I had gotten the information for the article and what my background was that I could author such an in-depth piece on electronic warfare. He was even more impressed that I had served in hostile situations in all three Navy elements, on, over, and under the sea. He had never met anyone who had made that claim. It was a most interesting lunch. Norman and I stayed in touch, and he quoted me in at least two of his later works. Thanks, Frank! I owe you a big one!

I met another man, Stu Harris, Robert Ballard's chief technologist during the hunt for the USS SCORPION, and then the Titanic. Stu had been the man who had developed the Argo, the remote vehicle which had spotted both ships. But as a submariner who had trailed Soviet submarines, I was much more interested in the results of the search for the SCORPION. This American submarine sank in the middle of the Atlantic under unknown circumstances. As a technologist, I was also immensely interested in the Argo and its forerunners. We had a great evening there in Newport talking about submarines and associated technology. I subsequently spent a couple of days at his home on

Cape Cod, but that is another story entirely.

This is just a tiny sampling of the many remarkable people I met at the NWC. I would have loved to stay there for my whole career. Indeed, I returned in 2000 as a civilian. If 9/11 had not intervened, I would probably have spent the rest of my professional life there. Still, I am getting ahead of the story yet again.

It was time for me to move on. Once again, I was recommended for a Legion of Merit. Once again, it was downgraded to a Meritorious Service Medal. Oh, well. I had heard this story before.

The head of Navy TENCAP, CAPT Ken Pelot, was interested in having me come to DC as his deputy. I was interested in going there, but my detailer saw I had no experience in a space ground station, which he felt was a primary requirement for the job, and refused to assign me there. It looked as if I was going to one of the large Navy offices in the Pentagon, either in Operations or Intelligence. However, one of my last trips from the War College took me back to USAF Electronic Security Command (ESC) HQ in San Antonio, where I saw many old friends.

Lieutenant Colonel (LtCol) Paul Martin and I had worked together when he was the PARPRO coordinator for the Pacific while I was at Atsugi with VQ-1. We worked together again when he had come to San Antonio as MGEN Larson's chief of staff as a new colonel about a year before I left to go to Newport. Paul, as I knew him, had continued to do very well for himself. He was now a Major General and the Commander of the ESC and the Joint Electronic Warfare Center, there in the same building. I stopped in to congratulate him on making not one but two stars in the four years since I had last seen him. He immediately offered me a job as a senior manager at the Joint Electronic Warfare Center (JEWC.) The Secretary of Defense had tasked the JEWC to write the nation's EW Master Plan. He was being given several billets along with the task. The team leader's job was still open, and he would like to have me there in that job.

I was delighted and accepted his offer. Then he asked what needed to be done with the Navy to make this happen. A sensible question, indeed! I gave him my detailer's phone number and suggested his aide phone the detailer and ask that question. Done deal! The rest is history. My old boss at Misawa, CAPT Pete Dillingham, was now a Rear Admiral and Director, Naval Security Group. He was delighted I had been offered this job.

So was Lloyd Feldmann, a young (mid-30s) GS-15 who worked on intelligence and space issues on the OPNAV staff. As we were setting up the Space Vulnerability meeting, he was one of my main points of contact. When he heard where I was going and what I was going to be doing, he asked me to stop by the Pentagon to meet three VADMs who he believed both I should meet, and they should meet me.

All three were most interested in getting a clear idea of our vulnerabilities in space and the opportunities EW gave us against the Soviet systems. All three tasked me with stirring the pot in San Antonio and getting these questions answered. None wanted to formally assign the JEWC as they might not be able to stand the answer, but they did want it answered and the results given to them, sotto voce.

Thus, I arrived at the JEWC with significant, if unofficial, tasking. The lack of official tasking was to cause me some amount of trouble and led directly to my decision to retire. But I left Newport with both pride and sadness in my heart. I was enormously proud of the job I had done and knew my bosses were delighted, too, but sad that there was still so much more to be done. I was excited to be going back to San Antonio for what promised to be a fascinating job.

Coda

I did see RADM Betterton again 16 months later. By then, I

was stationed at the JEWC, but I was invited back to the Naval War College in the summer of 1987 to attend Global War Game 87 (GWG 87) as a subject matter expert. There was an "icebreaker" reception the night before game-start in the Newport Officers Club's largest room. As I walked into the party, I saw him across the room talking with another Admiral. I did an immediate hard right turn and moved to the far side of the room to try to avoid him, but a few minutes later, he spied me across the room and waved me over.

I approached him with dread in my heart. However, His words cheered me up very much.

"Guy, I was hoping to see you here! You know I disagreed with your analysis the last time we met, but after that conference, I decided to spend $350K for an independent study to check it out. That study says you had, if anything, understated the problem, and I want you to know we are spending millions on fixing it. Can I buy you dinner tonight!

And that was the **High Point of my Career**, But I had already decided to retire at least in part due to his remark 16 months earlier....

That dinner was also my only reward for five years of study, other than my self-satisfaction, and that was more than enough. **My time in the Navy had been SUPER!**

The Joint Electronic Warfare Center (JEWC) Last Call!

Chapter Twelve

I hated to leave the Naval War College (NWC), where I knew I was doing significant work, which was also fascinating. But it was finally time for me to transfer if I was going to get promoted. I needed a headquarters-level job to complete my career qualifications for O-6, Navy Captain, my next rank.

At the NWC's Center for War Gaming, I worked closely with the several services' offices for "Tactical Exploitation of National Capabilities" (TENCAP). This experience led me to be asked by the head of Navy TENCAP, CAPT Kent Pelot, my ex-detailer (career advisor and assignments officer), to come to DC to be his deputy. However, our current detailer wanted someone with Classic Wizard (Navy Classified Space) ground station experience. The detailer thought it was a critical requirement in that job. The detailer just did not seem to realize that there was more to the military space program than the part for which the NavSecGru was directly responsible, which was the Classic Wizard program. The job with TENCAP was much broader than just Classic Wizard. It intrigued me as it would be a natural extension of my time at the Naval War College and my airborne time with the EP-3E and Rivet Joint. The detailer was dead wrong, but he held the hammer, so I was looking at going into some other slot, most probably in the Pentagon, but also possibly at NSA at Ft Meade or ONI in Suitland.

As I related in the previous chapter, not long after, I was told by my detailer that I was definitely not going to be assigned to Navy TENCAP, I attended an Old Crows conference in San

Antonio at my old USAF command, now called the Electronic Security Command (ESC). It also co-hosted the Joint Electronic Warfare Center. During a lull in the conference, I dropped in on the ESC's commanding general, MGEN Paul Martin, whom I had known since my time as Officer-in-Charge at NavSecGru Det, Atsugi. He was then a lieutenant colonel and the Pacific Command scheduling officer for all reconnaissance operations in the Pacific. We also worked together a second time during my operational evaluation testing of the RC-135W and its mission processing system for the USAF. During my last year with the ESC (1980-1981), then-Colonel Paul Martin had been the Chief of Staff for MGen Doyle Larson, so we knew each other well.

We had a very cordial conversation, especially after he opened it with, "How can we get you back here to San Antonio? We need you at the Joint Electronic Warfare Center (JEWC) to head our national EW needs analysis program we will be standing up next year." In that the JEWC was part of the Joint Chiefs of Staff (JCS) office, a very prestigious position, I was all for it. I suggested the most effective way would be for his aide to call my detailer directly to discuss how it could be done. The detailer was key. I had just lost one interesting prospect of a job because he could not see how I would fit into the TENCAP office. I knew he would have to be convinced by someone other than me.

That conversation between General Martin's aide and my detailer was held. It was agreed I would be assigned to the JEWC. There was indeed an empty Navy O-5 slot already there, and I did have the required experience. (Not a surprise!) The detailer even saw it as a good fit as that assignment counted as an assignment to the Joint Staff as we were an element of JCS-J3 (Operations). And, being on the Joint Staff was an excellent step up.

The JEWC had been created in 1979 by Major General (MGen) Doyle Larson while I was working for him. He was an innovator, having created the RC-135M Rivet Card COMINT collection aircraft and then the Rivet Joint (RC-135 V/W) SIGINT collection aircraft, among many other things.

He also saw that all services needed to pool their resources in the EW field as they were all facing the same threats and challenges. Trying to solve them individually was just plain silly and very wasteful. I was delighted to be assigned to such a vibrant organization, but I was to be severely disappointed.

I was first ordered to make a brief three-day stop in DC. It was requested by Lloyd Feldman, a young, very dynamic senior civilian on the Chief of Naval Operations staff. I had had many discussions with him when he attended various meetings and war games at Newport. Lloyd was in his mid-30's, and I believe he was one of, if not the, youngest GS-15s in history. He was assigned as the policy advisor to the electronic warfare section of the CNO's staff. He shared my concern about the US's lack of preparedness for strategic electronic warfare, especially against space systems, both military and civilian. Lloyd asked my detailer to give me temporary duty enroute my permanent change of station orders, not an unusual request. He wanted me to meet with him and several Admirals on the CNO's staff.

He wanted me to discuss how the JEWC could help analyze the threat of potential enemy's strategic EW capabilities with these senior officers. He also wanted the JEWC to study how these actions might affect naval operations. He also wanted me to understand precisely how concerned the Navy staff was in this regard, understand precisely what their concerns were, and take this message to the JEWC.

I first met with several one- and two-stars Admirals. These meetings prepared me for my appointments with all three VADM "warfare czars," the heads of the surface, submarine, and air communities. I had met with two of the three "czars" (surface and sub, but not the czar for Air) when they had visited the NWC's Center for War Gaming. Still, it is an entirely different matter to meet Admirals one-on-one in their offices. They have focused explicitly on you and you alone, telling you their specific concerns and what they want you to do about them. I left DC for Texas electrified, to say the least. I had a clear mission, and it aligned with my concerns, so I believed I had been anointed. I was severely disappointed here, too.

I arrived at the JEWC in early August 1986. I was assigned to be the deputy to an Army colonel with a human intelligence background. He headed the EW operational analysis division. His name is not essential to my story. He is probably dead now, so I do not want to distress his family. Still, even before I met him, I was told his men considered him mentally unbalanced. Subsequent events clearly showed me he was.

However, he and I got along very well initially. I had eight personal awards, an unusual number for a Navy commander in those days, and a total of eight rows of ribbons, counting campaign and unit awards, more ribbons than any other person, officer or enlisted, there at the JEWC. I was also wearing both Navy and Air Force wings. I am sure that gave me some credibility with him.

All went well for the first six weeks or so until one day, he and I were in his office alone. He asked me a question. It was something I knew a good deal about, and I gave him a detailed answer. The discussion proceeded normally for the next few minutes, and it was clear he understood my answer. However, he suddenly got furious and said something to the effect of making fun of him, taking him for a fool. "That is not what I asked you! I asked you...." and HE REPEATED HIS QUESTION of several minutes before. I looked at him in near bewilderment. But I had heard enough stories of precisely this kind of behavior by eavesdropping on the men in our unit that I was somewhat prepared. I repeated my previous answer pretty much verbatim. He said, "That's better." And we repeated the same discussion we had just. In the next several weeks, that fundamentally bizarre event repeatedly happened both with me and with the men of our unit. I noticed that when he was with anyone senior, other colonels, or generals, he was very ordinary. Still, he seemed to be struggling to contain himself. When he was with only people junior to him, his actions were often patently irrational. It was the joke of the office, but I kept my mouth shut.

I confided my observations to only one other person, CWO Monroe Lane, USN, the only other naval officer in the same division. We both agreed the Colonel was apparently not in his right mind a large part of the time. It was clear he was in control only when in the presence of seniors, the general, other colonels, or the Navy Captain, but not when with anyone else. When with men, he was senior to a completely different person and not in control of himself. It was both bizarre and concerning, made more so because he held a high clearance.

After a lot of thought, I finally decided that it was my duty to bring this to his seniors' attention. There were four O-6s, one for each service, colonels for the Marines, Army, and Air Force, and a Captain for the Navy. Each service also had at least one O-5, a Navy commander such as me, or a lieutenant colonel. I was, as a senior O-5, the second senior Navy man assigned to the JEWC. I should be coming up for selection to 0-6, Navy Captain in two years. MGen Paul Martin was the commander of the JEWC, but the assignment as the deputy commander rotated through the services O-6s. The Navy captain was the deputy at that time, so I decided to go to him. I had already given him the report of my stop in DC. However, the JEWC only had so many people assigned. It had other, more immediate, if not higher priority tasking. The requested in-depth analysis of the US's readiness for strategic, especially counter space, EW was, much to my concern, on a back burner. The captain and I had discussed this at some length, so I was very comfortable asking for an appointment to discuss a personal matter (the mental instability of the Army colonel).

At the appointed time, I arrived at his office and laid out what I had seen over the past month. I suggested the Army colonel needed to be ordered for a medical examination, including a mental evaluation. Something was not right. I stressed I had never done anything remotely like this in my career. I even offered to give him the names and phone numbers of all my previous reporting seniors.

I also stressed I was doing it only because the episodes of non-rationality appeared to be happening more frequently and with more violence. The Army Colonel seemed to be losing his touch with reality, and all the men in his division knew it. It was affecting the morale of both my division and the entire Army element at the JEWC. The captain said he would investigate what could be done. I left, believing I had done my duty and appropriate action would be taken.

What happened next was totally unexpected and still, 30+ years later, takes my breath away with its just plain stupidity. The Navy captain went to the Army colonel and told him my report in detail! He did not go to anyone else to the best of my knowledge except the USAF colonel. He asked him to accept me as his deputy, saying the Army colonel and I had a personality conflict! Bizarre! I did not have anything of the sort until that moment, but things got ugly in a heartbeat. My position at the JEWC was utterly undermined.

When the JEWC finally got permission and funding from the JCS to stand up the national EW analysis office, the job I came to the JEWC to create, the O-6 leaders gave the task to a highly capable fellow Navy O-5. He had recently arrived at the JEWC after having just completed his Ph.D. in mathematical modeling at the Naval Postgraduate school. I did go to work for the USAF colonel, but he was clearly out of his element. He oversaw technical vulnerability analysis of both friendly and hostile electronic systems. If he knew anything about analytical work, it was not evident, and all of us who worked for him knew it. Nice enough guy, but he was just putting in his time until retirement.

The analysis directorate had two divisions: Communications Systems vulnerability and Radar Systems vulnerability. Both divisions were further divided into offensive and defensive elements. The offensive offices looked at foreign systems to determine how they could be exploited, deceived, or denied. The defensive side looked at how to defend our systems from exploitation, deception, or denial. I was given the Communications Analysis Division, and that suited me just fine.

It was an enormous task and directly related to the tasking I had received from the CNO's office. However, we had other, "higher priority" tasking and only so many men, plus I had my tasking from the Navy staff only verbally, not in writing, at least initially, so I could not undertake the tasking the three vice admirals in the CNO's office had asked me to do.

I did report back to Lloyd Feldman, saying a tasking message from the Navy staff would be a big help. He may have been part of the instigation for a tasking letter we received a short time later via classified channels. It was from General Piotrowski, the USAF four-star commander of the US Space Command. However, its wording was unfortunate. Its language was remarkably similar to my report of my meetings in DC enroute to the JEWC and the recommendations I had made in that report! I think it was just a coincidence, but the similarity was not lost on the cabal of colonels at the JEWC. Not long after we received this tasking, I overheard the Army and Air Force colonels discussing the letter and deciding to ignore a four-star general's request. They were sure it was not really from him, but rather from me!

They thought it was from some Navy friend of mine on the General's staff that I had asked to send it as if it came from the general himself. (I am not making this up!) I did have friends there. And, we had indeed discussed this topic. However, my friends at Space Command had told me that both the General and his vice commander, VADM Doherty, were very interested in knowing all space systems' vulnerability, both ours and theirs. This is such an obvious need today that it is probably hard to believe that most of our officers once thought space systems were invulnerable to electronic warfare attack.

From my previous study at the NWC, I was sure they were vulnerable and that the Russians planned to attack them. Still, I was considered to be a fool by these two colonels for suggesting they were. I suggested to CAPT Rich Affeld, the deputy to VADM Doherty and a VQ-1 shipmate of mine, that he might want to have the VADM phone MGEN Martin and ask for a status report on the requested analysis. Instead, CAPT Affeld phoned the Navy Captain at the JEWC and expressed his concerns. This seemed to confirm to the Army and Air Force colonels that it was all a Navy effort that I had instigated, not really from the four-star himself, as it was.

Not long after that, I did attend the initial Space war planner's conference, held in Colorado Springs. None of the other attendees had ever participated in a war game on space war. I had actually led several and had also studied the subject in detail for over five years at that point, so I had a lot to say. But much to my disappointment, the conference's upshot was that the US needed to study the subject more. I had hoped for a more proactive outcome.

It was a waste of time, except for the sidebar discussion I detailed in my chapter on my time in submarines. Sometime later, well after I had retired, a space war game was held at Colorado Springs, and I read with some amusement the USAF report that it was "the first space war game" ever. Many folks who knew me also knew that was just not true, but I just laughed.

I continued my duties for another year-plus working on plans to defeat several countries' C3 systems, including North Korea and Iraq. Then, out of the clear, I was offered an outstanding job at an exceptional salary. Even better, it was in Newport, where I still had a lovely home. I had just learned that promotions to Captain for all officers in my specialty had been delayed at least a year, and probably two or even three.

This delay was caused by an order from the Secretary of the Navy. Congress sets the number of 0-5s who can be promoted to 0-6 every year. The Navy gets a specific number of those 0-6 promotions every year. The SecNav order significantly reduced the numbers of 0-6 advancements in the Intelligence, Medical Service, Cryptology, and Meteorology communities. Their promotions numbers were reassigned to naval flight officers (bombardier/navigators (B/N)) from the A-6 fighter-bomber community.

The Secretary of the Navy, John Lehman, was a Navy Reserve officer. His Navy career specialty was, you guessed it, A-6 B/N! This meant that I would not be in the promotion zone for O-6 for at least one more year, maybe two or even three. The delay in promotion was a bit of a demotivator for many of my year group. Indeed, all three of the men I had considered my "competition" for admiral, decided to retire as soon as possible. The SecNav's order also meant that I would continue to have to put up with the disparagement I was publicly handed near-daily by the Army colonel. I had hoped I would make 0-6, and the Army colonel would cease his daily public attempts to belittle me.

This ruling by the Secretary of the Navy was the final straw beyond the continual harassment from the Army colonel, which I tried to ignore. Being belittled in front of the troops was not my idea of fun. In hindsight, I should have gone to MGEN Martin about two months after I got there.

One nice thing happened just before I left. We had an "all-hands" party at a huge, typical Texan barbecue place. Toward the end of the party, the Army Sergeant Major, the senior enlisted man in the Army element of the JEWC, came up to me while I was standing off to the side watching the dancers. He held out his hand, and I took it. As we shook hands there at the barbeque place, he gave me one of the nicest compliments I ever received. It was to the effect that I had the respect of every Army man in the command, officer, and enlisted, for my attempt to handle the Army colonel problem and the way I had dealt with the situation afterward. Monroe Lane must have leaked my efforts to get at least the Colonel's top-secret clearance removed, as I am sure I had told no one other than him what I had recommended to the

Navy Captain. And I doubt the Captain had told anyone, but maybe he had.

The Sergeant-Major went on to say that he had over 30 years in the Army, and the Colonel was the oddest thing he, or any of the Army men, had ever seen. It was a nice affirmation that my judgment was correct. It was a weird situation. All I could say was, "Thank you, Sergeant Major. I appreciate you saying something. Let's enjoy the party!' And we did!

My retirement ceremony was not without its drama as well. I phoned Admiral Stansfield Turner, the man responsible for my career in the Navy, to tell him of my decision to retire and thank him for making it all possible. He offered to be the officiating officer, a true honor, if I could find a time that worked with his schedule. He gave me the phone number of his aide/assistant and told me to call him to see what could be worked out. I did, and we determined that 19 September worked best for the Admiral.

I went to the Protocol Office, which controlled the auditorium where I wanted to hold the ceremony, and asked to reserve it. They asked why and I told them it was the only time the presiding officer, a Navy Admiral, had a free spot on his calendar. They informed me that it was the Air Force's tradition that retirements were celebrated on the last Friday of the month you retire. I pointed out that I was a naval officer, and it was a Navy tradition that you could retire on any damned day you wished.

They were not pleased but agreed I could have the room on the date and time I requested, but they stipulated that they would provide no support. No printing of the program, no mailing of invitations, no catering, no room setup, no assistance with guest parking, etc. I have organized large parties before, so I was not concerned.

A couple of weeks later, my Colonel came to me and asked why I insisted on September 19. He informed me that Protocol was angry with me. I told him it was because that was the only date that the officiating officer, who had been a significant part of my career, was free.

A few days later, the Colonel told me General Martin was curious as to what was the rank of my officiating officer. "Admiral," I answered. Being that General Martin and I were friends, I am sure he was curious why I had not asked him. In hindsight, I should have told him on day one.

The next day the Colonel asked me how many stars my officiating Admiral had. "The usual, four," I answered. I think the Colonel had assumed I was talking about a one- or two-star Admiral when I had said "Admiral" the first time. He was obviously more than a bit surprised when I said "Four."

The next day the Colonel called me to his office, and this time his question was much more direct.

"The General wants to know what Navy 4-star will be visiting the headquarters of Air Force technical intelligence!"

"Admiral Stansfield Turner."

"The Director of the CIA?" (His voice was nearly an octave higher.)

"Yes, Sir. But he has retired from that job."

And I left the Colonel's office and went back to my own, trying hard not to laugh out loud.

A minute or so later, the General's Chief of Staff, an "up and coming" colonel with whom I had cordially interacted several times, arrived at my door. He immediately and forcefully asked me: "Thomas, the immediate past Director of the CIA, is coming to Air Force Technical Intelligence Headquarters at your invitation. And you have not thought to alert anyone? What the xxxx were you thinking?"

"That you guys do not ask many questions, Sir. I did not want to flaunt it." I replied as calmly as I could. (But I was still laughing inside!) He graciously agreed with my reasoning, saying, OK, I get it! And we both had a good laugh.

Needless to say, I suddenly had more help from Protocol than I could use, and my retirement audience filled the auditorium. The crowd was several times the number of folks I had invited, and the rumors flew as to how I knew the Director, CIA. Of course, my reply to all who inquired was much along the lines of "If I told you, I would have to kill you." And then we would both laugh, and I would tell them a bit of the real story. It was much more prosaic than most guessed, but I did not elaborate. It did give me the last laugh.

Codas

About two years after I had retired, I ran into the Navy Captain from the JEWC at an Old Crows convention in Dallas. (The Old Crows are the professional association for electronic warfare officers.)

He introduced me to the man with him, a senior vice president (SVP) of Lockheed Martin, the company for which he was now working. Without saying a word about our past, the Captain recounted the story above about the crazy Army and ignorant USAF Colonels to the SVP.

He even explained how his telling the Army Colonel that I had recommended removing his clearance was probably the stupidest thing he had ever done. He also said I had behaved like a true gentleman. Still, his actions had likely forced me out of the Navy, and I was one of the most competent officers he had ever served with. It was an unequivocal and heartfelt apology, and I forgave him. We all do stupid things occasionally in our life. Still, it was the turning point in my career, which had been pretty much straight up until that point. I ran into the Captain several more times in the following years, and he always repeated his apology. It almost became a ritual. In some ways, he did me a favor as I got an excellent job upon retirement. Yet another blessing in disguise?

****

CODA 2

About four years after the above meeting, I ran into MGEN Paul Martin at another Old Crows convention, this time in Colorado Springs. He had just retired from the USAF and was now a consultant. He spied me in the bar at the convention hotel the night before the convention started and came over to say hello. He joined me, having a coffee to my beer. After we had caught up a bit, I asked whatever had happened to the USAF Colonel I had worked for.

"Stupidest officer I ever served with!"

"What about his running mate, the Army Colonel? I asked.

"You know, Guy, he was as crazy as a loon. I could not stand talking with either of those men. My secretaries had standing instructions that if either of those men wanted to see me, I would be too busy, even if I was just sitting in my office reading the Wall Street Journal."

I followed up with: "What do you think it would be like to be working for either man?

The general replied: "I could not have done it. I would have gone postal and killed them!"

My response: "Do you remember who I worked directly for when I worked for you at the JEWC?"

"Them! How did you stand it?"

"Why do you think I retired when I did?"

"Jeez, Guy, I wish you had come to me! We could have worked something out."

"I did not think it was appropriate, but I probably should have. And being offered out of the blue to be paid as a 3-star and relocated to Newport, RI, where I wanted to retire anyway were also powerful motivational factors and gave me an unbelievably soft landing."

"Yeah, I remember you got an outstanding job. Best I ever heard of out of San Antonio! You were making more than me!

So, it was not just me! I always knew it was not, but it was both good and bad that the general was aware of those two men's shortcomings. It was affirming, but they both should have been forced to retire long before I did.

****

Coda 3

In the late 1990's General Piotrowski, the man who had commanded the Space Command and sent the tasking letter to the JEWC, now retired, and I met at a meeting at JHU/APL.

We were at a day-long briefing on where the US was in the current Space Domain race. I asked him whether he remembered tasking the JEWC to do a vulnerability analysis of space systems. He clearly did remember and expressed his dissatisfaction that it was not accomplished during his tenure. He asked me if I knew why it had been delayed.

I certainly did and told him why. He was incredulous. Then I told him the reaction of the JEWC to his tasking had caused me to retire. He, too, expressed his regret that I had not gotten in touch with his staff in a more forceful manner. I now regret that, also.

Finale

My first two years in the Navy were not all that much fun, I was not making much money, but it was very instructional. The same can be said of the last two, but, sadly, they were even less fun. Still, all in all, I would do it all over again, even if I were so wealthy that I did not have to work. I might have lived in a better house, driven a better car, and traveled more, staying at better hotels, but the Navy was not a job. It was a privilege and a calling.

This was also true of all the Armed Forces of the United States. I worked with all five branches, including directly for the Coast Guard as a civilian. Every service is full of dedicated men and women. My calling was to the Navy, but I always respected my fellow servicemen, officers, and enlisted of every branch. Still do.

Besides my time in the Navy and with the US Air Force, I also spent nine years with the US Coast Guard as a civilian when I was the US' Science & Technology Advisor, basically, the Chief Technology Officer of the Maritime Domain Awareness Program Integration Office. I am very thankful I had the opportunity to serve with three branches of our outstanding military! ¡Gracias de Dios!

Newport to Dallas to Pax River to APL A Dog's Breakfast!
Chapter Thirteen

Besides my brief stint with the FBI, which I will detail in the next chapter, the years between my retirement from the Navy and 9/11, my life was widely varied, pretty much a "dog's breakfast" of jobs. They were often on the leading edge of technology and thus highly classified. Indeed, I am not sure what I can talk about even all these years later, so I will only go over them lightly. It was all fascinating, and I continued to learn a great deal.

From the JEWC in San Antonio, I returned to Newport to run the Northeast region for Information Systems and Networks (ISN). I replaced a retired USAF Major General who had decided to retire fully. I was the 9th in seniority in the company and was paid as a three-star. We maintained the local area network (LAN) for the Naval Underwater Warfare Center and its two satellite offices, one in Groton, CT, and the other on Andros Island in the Bahamas.

While I was there, we won the contract to install the LAN at the Naval War College. It was a fun gig, but the two ladies who ran the business were micro-managers of the first order. When six of the top 12 officers of ISN, a 650-man organization, resigned in a noticeably short period, I decided they knew something I did not. I quickly accepted the standing offer from E-Systems (ESY), the builders of special mission aircraft, including the USAF RC-135 fleet (Rivet Joint, Cobra Ball, Combat Sent), plus a few others, the most famous being Air Force One. It meant a return to Texas, but the offered pay raise and the lack of a state income tax were enough to make me leave beautiful Newport.

They also provided me the chance to return to Japan as part

of their team working on replacing many of the antennas on AF-1 with the first multi-band phased-array communications antenna ever placed on an aircraft. It sounded like fun. It started as fun, but the negotiations dragged on for over two years. One nice thing was that we were flown to Japan in First Class and stayed at the best hotels in Tokyo. It was a significant step up from travel even with the Air Force, much less the NAVY! I logged almost 600,000 miles on American Airlines alone in those 28 months.

The Japanese also wanted E-Systems to build them a copy of Air Force One, which the Texans at ESY were all for, but then the Japanese made it clear they wanted it built in Japan. Whoa, there, Nellie! That was a whole different kettle of fish. We spent a lot of time in DC talking to the State, Commerce, and Defense Departments. State and Commerce liked the idea, but Defense was dead set against it. Shortly after Defense won the day and decreed E-Systems would not be building an AF-1 for the Japanese, the company decided it needed to cut back its workforce. Over the next six months, it laid off 6,300 of its almost 19,000 employees.

They were primarily concentrated on those of us over 45 years old and making over $50,000 who were not working on a specific, funded task. I was vulnerable on all three counts but was still one of the last folks laid off. The following 22 months were a lesson in humility.

While working on the Japanese/AF-1 project, I had also been working on a counter-narcotics project, studying the opportunities for ESY in the "War on Drugs." I spent a significant bit of time on the southern border out of El Paso and at Key West, Riverside, and Oakland, then the four centers coordinating border surveillance. El Paso was doing land surveillance, Riverside the air, with Key West doing the maritime watch over the Caribbean, Gulf of Mexico, and Atlantic, while Oakland had the Pacific.

After studying the problem for almost a year, I reluctantly realized that there was no real money to be made in that business, and my report said just that. It was a waste of time and money. Judging by the money being allocated and how it was being spent, the prize was not worth the effort. However, the time I spent there stood me in excellent stead when I went to work on the maritime domain awareness problem in 2002, 12 years later.

I did regain my access to special intelligence, but only after a several-month lull, the reason for which I will describe in detail at the end of this chapter. I subsequently attended many classified briefings, both there at Greenville and other locations. At one of those briefings at Greenville, it was announced that one of the scientist-engineers had developed two different PCMCIA cards with the library of how to identify and also how to record all known video signals of "special" interest on one card and all known audio special signals on the other. Not long before this brief, we had also been briefed that the GRIDCase computer company had just developed a "belly-pan" for its highly regarded GRIDCase ruggedized laptop, the first laptops in space, having flown on the first Shuttle flights.

The 'belly-pan" had two PCMCIA card slots and the ability to exchange data between them and the host laptop. The idea immediately occurred to me that this would be a handy tool, especially on a submarine. I approached the inventor and asked if he had considered taking these new cards to sea. Greenville was an Air Force installation, and he had not, but when I described how it might be used at sea, he was intrigued.

In the next month plus we ran a series of tests on the idea, which was not without its problems due to time-lapses between the cards in the belly-pan and the primary software resident in the laptop itself, but we were able to figure out how to make it work by segregating the library functions from the recording functions. It was an elegant tool that I would have been delighted to have with me in a submarine, ship, or plane off a hostile coast. My idea was to segregate the two functions that allowed the device to perform as required, so I had some pride of ownership.

I made an appointment to visit the NavSecGru Direct Support shop in the NSG Headquarters and lugged the GRIDCase to DC. Their reaction was mixed, but all agreed it would be extremely useful. The problem was that they wanted me to give it to them for free. To make a long story short, we gave it to them for a trial run onboard a special mission submarine, where it earned rave reviews from the embarked team. Indeed, they wanted to keep it, but E-Systems wanted it back for more development, plus E-Systems wanted a contract to build a number of the units to put in the fleet all over the world.

I thought the price was very reasonable ($25K a copy), but NSG balked at it, and then I was out of a job and lost track of my device. However, several years later, after I had regained a position that required that level of clearance, I happened to attend a briefing at the Office of Naval Intelligence. Several of us NSG types and a few others ended up at the same large table at lunch. One of the officers there was the man from NSG HQ who would not buy my device, but I did not mention it.

However, toward the end of lunch, he expounded about how he had come up with the idea for a device that did exactly what I had tried to sell him. It was now deployed in the fleet to rave reviews! He used a slightly newer base computer with the shape of an old-fashioned, round-topped "lunch box" and called it the "Lunch Box Special Processor." Still, its functionality was precisely the same as my GRIDCase special processor. I suspect they pirated (AKA STOLE) the software off my GRIDcase and ported it onto the Lunch Box Special Processor.

If he had given me a contract to bring my device online and deploy it, I might not have been laid off and spent 22 months under-employed! It also really irked me that he was claiming all credit for something I had conceived and spent some time helping the smart guys at Greenville build and perfect. I nearly threw my lunch tray at him that day at ONI, except he was at the far end, and there were about ten people between us. This would not be the last time someone claimed credit for my work, but maybe the most irritating! I hope his colleagues will read this and realize where he got the idea for the "Lunch Box Special Processor."

A couple of months after NavSecGru had declined to buy my "GRIDCase Special Processor," my whole division at work, about 50 people, all with at least one advanced degree, and most, including me, with two, were laid off. We were given three weeks warning that we would be laid off. During that period, I was offered a job at the plant for about 2/3s of what I had been making and foolishly turned it down. In hindsight, I should have taken that job and then, while employed, kept looking for another at near the old salary! Big mistake! HUGE! Very costly! Live AND Learn!

I spent the next 22 months looking for a job and spending a significant amount of time doing volunteer work for the Dallas Opera. At least the work at the Opera was great fun. I had many roles there, including working performances as a salesman in the boutique. I was subsequently asked to run all aspects of lobby operations several times when the Cowboys were in the NFL playoffs. The games were on the same day as an opera. Many of the staff went to the games or parties, and I stayed and ran the "front of the house." I felt honored to be entrusted with that responsibility, especially since I was a volunteer.

I was also a photographer at money-raising events. In addition, I was the assistant treasurer, keeping the books for the Opera Guild. I helped set up their books using Excel spreadsheets, then a new tool. I was also co-chair of Acapella, the singles organization of the Opera.

My partner, Penny Plueckhahn, and I took its membership from 200 lackadaisical members to over 800 active members in that role. Acapella had both great parties for arts-loving singles in the Dallas area and provided very substantial workforce assistance to the Dallas Opera staff. We subsequently won an international prize as the best volunteer organization supporting an arts organization in the world. That award also got us a very nice article in the Dallas Morning News. It is another award I am immensely proud to have won. Like nearly all my other awards, it was a team effort. Penny was a marketing genius. Literally a genius. That was her profession, and she was obviously exceptionally good at it. I knew how to use all Microsoft Office tools, a rare skill in those bygone days. We were a great team!

After 22 months of under-employment, I took a job that required me to move to Patuxent River, Maryland. I was part of the team coding the simulators at the Air Combat Test and Evaluation Center (ACTEF). They hired me to be part of the "Red Team," I worked on modeling Soviet capabilities and tactics. While in that job, I somehow got the nod to brief a professional conference in Colorado Springs on the capabilities and tasks underway at the ACTEF. At the conference, I ran into several people from Johns Hopkins' Applied Physics Lab (JHU/APL). I had worked with two of them at the Naval War College's Center for War Gaming in Newport nearly ten years earlier.

They expressed surprise that I was not still in the Navy. They then said that I should have applied for a job with them. They were always looking for folks with real-world military operational experience which also understood technology. One of them indicated that he was sure there was an open "requisition" at APL for someone with my exact experience. He urged me to apply for it immediately. I went back to my hotel room that night and did just that. APL was the best of the best to my mind in those days, and if I could get on there, I probably would never leave!

Two months later, I was working at what many considered to be the premier technical think tank/laboratory in the US Navy's system, if not the world. Its list of accomplishments in the technological world is storied, reaching back to before World War II. Its creation of the proximity fuse is almost unknown, but many experts consider it one of the three main reasons the allies won the war. The other two reasons are much better known, the

Atomic bomb and the Enigma machine. The proximity fuse allowed the allied anti-aircraft fire to be many times more lethal. All an antiaircraft artillery shell armed with a proximity fuse had to do was pass close enough to an aircraft to be sensed by the fuse. The fuse would automatically explode the shell at its closest point of approach. This technology was Top Secret for many years.

It is also not well known that APL was the creator of satellite navigation. Today most of us carry around smartphones with GPS, the off-spring of the device created at APL, without a thought that its forebearer was created to make our submarine-launched nuclear ballistic missiles more accurate.

My first job was to help build a Warfare Analysis Lab Exercise (WALEX) examining how best to counter ballistic missiles such as the SCUD used by the Iraqis. It had been used by them both during their war with us and their earlier war with Iran. We had three primary alternatives: upgrade the Army's Patriot anti-ballistic missile (ABM) capabilities, enhance the Navy's Standard missile system, or go with the Air Force's Airborne Laser (ABL).

We held a series of WALEX's as we went deeper and deeper into the subject, and I came to understand much more fully what a great tool the WALEX Process was. It was created there at APL by combining research with wargaming in an iterative fashion. I am still a big fan today. It is an excellent tool and learning what you did not know you did not know (sic).

WALEXs were excellent tools to help find ways, via the focused interaction of operators and researchers, to narrow the areas of no understanding. Then you test again to see what else you did not know you did not know. You repeat the process again and again until you know you understand all that is pertinent. Or at least until you know the right direction to move forward. This process allows one to concentrate their research on the right problem (s) at hand.

APL received a contract to play a substantial role in defining the sensor set and its related data links for the Joint Strike Fighter (JSF). I was picked to be a principal investigator on that project. As part of my task, I visited the Navy's TENCAP office. Several of my friends from my VQ-1 days and others from my Space desk days at the Naval War College were there. But what surprised me was seeing Riley Perdue, one of the bright young sergeants from my USAF exchange days, there and in a position of authority. Riley and I had hit it off when he was stationed on Okinawa as a flyer on the RC-135s. I visited there to fly familiarization flights and toured the squadron's operations and analysis spaces. We struck up a conversation when I was introduced to him as a guy who "knew his stuff" in an area I was particularly interested in learning more about. Over the next two years, every time I visited Okinawa, we had some great talks. He was a master of his trade and rightly proud of it. I was not the only one interested in this area, but those of us who were could have fit in a small school bus in those days. (Now it would take an NFL stadium). He even invited me to his E-5 promotion party. I was the only officer there. I felt truly honored, and I thought to bring a case of beer! Riley remembered that case of beer all those years ago. It was Coors!

Riley was there working on the Integrated Broadcast Service (IBS) program. It was the "child" of the Tactical Intelligence Broadcast Service (TIBS). TIBS was a direct result of the Rivet Joint Block III (RC-135W) test report that I had done in 1981, so I felt right at home.

Starting in about 1985, each service, Army, Navy, and Air Force, had developed its own tactical data link. Desert Storm had shown that we needed to combine these three tactical data links to provide an integrated picture. Each service had strong opinions that theirs was the best. Riley and his office mate, LCDR Brian Johnson, another NavSecGru guy, were trying to sort it all out. The three of us discussed how the task of integrating these several data links would be accomplished at some length. We also discussed how the information generated could be provided to the JSF in near-real-time (less than a second in military terms.). I included their thoughts with mine on how IBS could be integrated into the JSF in my report. It became an important part of the analysis of the C4I needs of the JSF that JHU/APL was writing.

As a courtesy, and to get their last thoughts, I showed my input to the JSF analysis report to them before I sent it up my chain of command. They liked what I said and how I said it. I was not surprised they liked it because a lot of what I was saying was their own words and thoughts, just reorganized, so they flowed naturally. However, it did come as a bit of a surprise when they asked me to come to work with them. After some juggling of folks at APL, I switched from the JSF team, which was winding down anyway, to being part of the TENCAP IBS effort.

It was like jumping from a frying pan into the fire. The master plan had already been written, but Riley and Brian had serious problems with it, and in reading it, I understood why. It was too focused on the Navy problem, while this needed to be a tool for all four services. The Marines had also joined the effort and were making their needs known as Marines do everything, forcefully and with passion. The program was again running into technical problems. My first task was to write the security manual for the effort, assigning the classification level of every aspect of the program and its product. I had the security manuals for each of the four systems we were combining/integrating. I combined all four security manuals into one large document. I then removed the duplication, organizing what remained into one smaller document. It was a lot easier said than done. Still, it was an excellent way to learn all the systems' different aspects and the overall problem.

Brian rotated out about then, and LCDR Kathy Helm took his place. She was a petite but very smart lady. She was also very focused, worked hard, and expected everyone else to do the same. I ended up re-writing the master plan, using much the same methodology I had used to create the security manual, but it was much more complicated. It became the guide for the rest of the effort.

Shortly after this, the USAF took over responsibility for the IBS effort, and I was asked to stay on. I was pleased to do so, but out of the blue, my boss called me in to discuss an unknown subject.

"Guy, you have indicated several times that APL should reestablish its liaison billet up at Newport and give that job to you. As you may know, we have had a young PhD there near full-time for the last six months. (I did not know that.) He has just asked to be relieved as he wants to stay down here full time with his young family. We would like you to move to Newport as soon as possible to relieve him. What do you say?"

"I think it is a great opportunity, and I am extremely interested, but I had better ask my wife first. You know we have just moved into a grand old Victorian mansion near downtown Baltimore. We have spent almost every waking hour for the last five months getting it ready to be a bed and breakfast she is planning to run."

Luckily, I had taken her to Newport the previous summer, and she had loved it, so getting her concurrence to move north to Newport for a couple of years was quite easy.

At Newport, I was assigned to the Naval Warfare Development Command (NWDC). It organized, ran, and analyzed the Navy's live battle exercise program called "Fleet Battle Experiments" (FBEs) as its primary mission.

APL also tasked me with keeping in contact with the rest of the Navy's research efforts in Newport, including those at the Naval Underwater Warfare Center and the other Naval War College's Center for Naval Warfare Studies besides NWDC. These included the Center for War Gaming, where I had previously worked, the Strategic Studies Group, which I had previously assisted, and the Center for Advanced Research, where

I had been assigned when I was pulled from the student body. My new job was a hectic full-time job, but I loved it.

Clelia and I also volunteered to sponsor an international student, specifying we spoke Spanish, so, of course, we got a Malaysian family. Come to find out, there was a waiting list of sponsors who spoke Spanish. We did not really care. The Malaysian officer and his family were charming, and we loved the multi-national parties.

I had already participated in several FBEs, testing new analysis tools while still at APL, so I both felt right at home and already knew a large percentage of the staff, including all the leaders, who welcomed me very warmly. Even my study partner of 1981/1982, then Commander, now VADM Art Cebrowski, the President of the Naval War College, made a point of warmly welcoming me publicly. It was an excellent way to start my tour. I became one of the analytical staff.

Then, very quickly, CDR Mark Chiccone, the game director for FBE India, to be held in the summer of 2001, 15 months in the future, asked me to be his technical advisor. Mark was an E2C naval flight officer. He had come straight from squadron command and was another guy with "Admiral" written all over him. However, shortly after his tour at the NWDC ended, and he returned to the fleet as the E2C wing commander, a big step toward Admiral, he experienced a deep vein thrombosis and was medically retired. It was the Navy's loss. He would have been a fine Admiral!

However, while we worked together, he was "healthy as a horse," and he ran me hard evaluating the many technologies which wished to be part of our experiment/exercise. In all, 254 technologies were presented to us, each formally requesting inclusion in our experiment. There was a team of seven of us who analyzed the requests. Thankfully, these requests came in a standardized format so we could compare "apples to apples" and "oranges to oranges."

The seven members of this panel were all retired Naval officers with technical backgrounds. They included at least one person from every warfare specialty in the Navy. If we saw a technology with which none of us were familiar, we used our contacts across the entire Navy to find someone who was. Between us and our ready connections, we had state-of-the-art for modern naval warfare well understood. In the end, we selected 54 technologies out of the 254 to participate, and we wrung them out hard during the actual FBE. It was a great learning experience for all involved. Especially for me. It was super preparation for my subsequent job as the Science & Technology Advisor in the Maritime Domain Awareness Program Integration Office. (MDA/PIO).

At the end of the FBE, I stayed behind for two more days to tie up some loose ends and share my draft report with some of the 3rd fleet staff. Their flagship had been our headquarters for the entire time, and I thought it only fair. They had helped us all a great deal during the Fleet Battle Experiment. I wanted to be sure none of them had any problem with what I said in my report, which was "The Good, The Bad, and The Ugly." Indeed, some of it was very ugly. But all agreed it was a fair treatment. But some contractors even got fired (and rightfully so!) because of my report.

I came home in mid-August 2001 after being gone for over 50% of the last seven months. I went to work the following Monday, and we had a command-wide debrief of the FBE at the end of the week. After that debrief, we were told to go home and not come back until the day after Labor Day. I remember I attended a briefing at ONI on the Passive Coherent Location (PCL) threat right after Labor Day and was busy writing a description of how I thought that threat should be addressed for the next day or two when 9/11 happened and my life took yet another sudden twist.

****

Before moving on to how satellite AIS came to be, and my getting very deeply involved in Maritime Domain Awareness, I should outline a very ugly situation in this otherwise normal period.

When I opened the ISN office in Newport, I was told to hire an administrative assistant. I chose ▮▮▮▮, the admin assistant in the NWC's Intelligence Office. She had experience handling classified information at higher classifications than anything we had in our office, so her experience would be a big help. Unfortunately, she had a series of mental episodes, including one where she mailed a large package of company proprietary information to one of our competitors. I found out about it when a friend at our competitors' office called me and told me they had received the package, and realizing what it was, had not gone beyond the first page. They brought it back to us.

I had already informed Mauri that her job was in serious jeopardy if she did not get medical and psychiatric attention. Still, she refused to acknowledge she had any problem, which everyone in our organization of about 15 people knew she did. They had all witnessed what can only be described as bizarre actions by her. It was the talk of the group, but I explicitly did not participate in it as I was still weighing my options. I confer with my two most senior men, one a retired naval aviator I had known for years. We all agreed it was time for her to go. I reported the situation, including the incident of the mailing of classified information, to my superior in the company, and it immediately went to the two sisters who owned the company. I was ordered to fire her immediately, and I did.

She walked out the door and went straight to the Naval Criminal Investigative Service, the NCIS now of TV fame. Her first statement to them really got their attention.

"Guy Thomas fired me because I discovered he is a spy! He was part of the Walker spy ring!"

This was not that long after the Johnny Walker spy ring had been uncovered. Many intelligence professionals believe Walker was the most damaging spy ever, and he indeed was regarding naval operations. NCIS was still twitchy from the aftershocks of that scandal and took accusations such as hers very, very seriously. Rightfully so, in my opinion, however, the next 18 months of my life were very unpleasant.

Many people I knew reported they had been visited by either NCIS or the FBI and questioned in detail about me. They all knew these questions went well beyond a routine background investigation. After a year of this, I filed a Freedom of Information request. I learned that my office and its safe had been opened and gone through. I also suspected I was under observation.

But it was the treatment of my family that angered me. My daughter, a sophomore at a College in San Antonio, reported she was 100% sure she was under observation. She even had a new student, who had arrived in mid-semester, befriend her for no apparent reason. Then the "new student" had brought up items from our past that clearly indicated the new "student" knew she had lived in Japan and New England. After several more weeks had brought up Walker and asked her if I had ever mentioned him, my daughter replied that I thought Walker should have been executed, which was true. However, the next question really did give the game away. "Your Dad (me) wanted Walker dead to silence him? Was he worried about being incriminated?"

No, my daughter told him I wanted him dead because I (me) thought, "He was a traitor of the worst sort. Walker did it for money. Even worse, American servicemen died because of his treason. Because he had given the encryption keys to some of our most sensitive codes to the Russians, and the Russians had given the resulting information to the North Vietnamese, they often knew our strike plans before the pilots who were going to carry them out did. Thus, the Vietnamese could concentrate their anti-aircraft resources where they would do the most damage to our Navy, Marine, and Air Force flyers. There is no way of quantifying the damage Walker did, but it was very substantial." My daughter had heard me expound on Walker enough times she knew my feelings about him by heart!

During this period, I moved from Rhode Island to Texas to work for E-Systems on classified projects. Before I left, I informed NCIS I was moving and what I would be doing. I also asked if they had any problem with me retaining my clearance, which would be necessary for my new job. They assured me they had no issues with me leaving and going to another cleared job. I was routinely debriefed of my Top Secret /SCI clearance; however, when I arrived at E-Systems, I discovered that I was blocked from being re-indoctrinated. This blocking was because, allegedly, the now six or seven-month-old case was still under active investigation.

This situation dragged on for about another nine months. Finally, the Defense Investigative Service (DIS) agent in Greenville, Texas, responsible for the E-Systems plant and its 6,000+ cleared individuals, including my case, decided to call Newport and see what they had on me and what the holdup was. He had interviewed me when I first arrived. That was his job, and he was curious why someone with my record, which he had total access to, was being held up in being re-cleared. Maybe I really was a spy! He did not actually think that, but he was curious. I had given him my point of contact there at NCIS Newport and his phone number when I first arrived the previous Fall, and now it was early summer, a year later, so he gave him a call.

He told me afterward that the response dumbfounded him. NCIS Newport had quickly decided that there was no merit in my ex-admin assistant's claim. They had "Just not gotten around to closing the case," which required a small amount of paperwork.

(This was while my life had been put on hold for over a year!)

The DIS agent asked his superiors in DC for permission to reinstate my clearance. They initially said yes, then changed their mind, probably because of my space systems and submarine operations clearances, both of which were very restricted in those days. The DIS agent told me his boss wanted me to take a lie detector test. Would I agree? I had been offering to take one since day one, but that was not in NCIS Newport's file. Indeed, they had told the DIS agent that I had declined to take one, which was an out-and-out lie. They were just covering up their laziness and incompetence.

I subsequently took a four-hour lie-detector test that seemed very thorough. At the end of it, the examiner and I walked out into the main office. He announced to the several agents in the office that they should buy me lunch as this was a clear case of a vengeful/vindictive woman. At that point, I was so paranoid that I thought it was some sort of a trick! I did enjoy lunch at a favorite Texas barbeque place of theirs.

I can still remember seeing the E-SYSTEMS Special Security Officer coming down the plant's main hall two days later. He spied me and boomed out: "Guy, The DIS folks phoned yesterday and told me you are as clean as a hound's tooth. Meant to call you and tell you to stop by the office to be re-briefed." I can still feel the weight lift off me when I think of that moment.

My Time with the FBI
Short but Very Sweet!

Chapter Fourteen

This part of my life goes back more than twenty years, to 1993-94. I believe that is enough time that I can now discuss it without jeopardizing national security, or the safety of a grand old man. If he is not dead now, he must be well over 100 years old. Let me begin at the beginning of this adventure.

E-Systems was a premier aviation electronics firm that took existing airframes and made them very special. Its crown jewel was Air Force One, but they also manufactured the Rivet Joint, Combat Sent, Cobra Ball, and Combat Talon reconnaissance/special mission aircraft. They also built numerous head-of-state aircraft for countries all over the world.

In 1989, E-Systems hired me to be a significant part of their effort to sell a copy of Air Force One to Japan. In early 1992, the United States government decided it was not in their best interests to sell copies of the technology in Air Force ONE to anyone. We were forced to abandon the effort. To make matters worse for me, the Soviet Union had fallen apart, and the Berlin Wall had come down. Most people believed everlasting peace was at hand. Many thousands of defense workers were let go worldwide, even in the United States. E-Systems, a company with nearly 19,000 workers, laid off about a third of its workers. Many other defense companies all over the nation laid off similar percentages. I was one of those let go.

I used my newly imposed free time to volunteer for several music organizations, especially the Dallas Opera and the Dallas

Summer Musicals. Because I could carry on a conversation with anyone from a yardman to a senator, I was routinely invited to various fund-raising parties to fill out the crowd and help talk up the organization of the day to raise money.

One of my fellow volunteers, an attractive lady who was a senior administrator in the Dallas public schools, happened to bring a date to one of the parties in mid-1993 after I had been out of work for over a year. During our conservation, her date, Richard, learned I had studied Russian and became quite interested. I gave him my usual cover story about liking languages and studying Russian, hoping to be an interpreter at the 1980 Olympics. He did not ask any more questions about how I came to know Russian. However, he was looking for someone to act as a Russian interpreter for him. He was an entrepreneur with several business interests. Texas seems to have more than its share of the type of men who dream big. Some make it; many do not. I judged this fellow as one who just might succeed. He was, at about 45, still in there swinging, as they say.

I had no job at that point. I, and over 6,000 of my closest friends from E-Systems, had been laid off the previous summer. Thus, I was willing to listen to any proposal that was legal. It turned out the small office building owner where Richard had his office was an émigré Russian of Iranian descent. He had an older brother who had recently arrived in the US. His older brother was a dual doctor with an MD in radiology and a Ph.D. in nuclear science. He spoke nine languages but no English. He spoke Russian and five central Asia countries' languages, plus Iranian, Arabic, and Turkish. His lack of English meant he was unemployable in the US, so he wanted to set up a trading company between the US and contacts he had in those central Asian republics of the former Soviet Union. He was also a Muslim Iman in the Shiite sect but had been brought up in a Sunni area and still well connected in Iran and several other countries in that part of the world. I subsequently learned he had taught in several major universities in most, if not all of the 'stans...Pakistan, Afghanistan, Uzbekistan, Kazakhstan, etc. He had been given a tourist visa, allegedly because he had family in the US. He brought his wife, about 70, and his daughter and her husband, both in their late 20's.

The initial interview was quite interesting. I met Richard at his office. After a few minutes of small talk, where I reiterate my Russian was quite rusty, he excused himself. The doctor walked in a minute later. He was in his middle 70s but still quite active. He introduced himself in Russian, and we talked for about 15 minutes, with me stumbling along and apologizing for my rustiness. He then excused himself and left the room, and Richard returned a minute later and said, "He said you would do. You have a job." It was only a part-time job and did not pay much, but at least it was something.

The doctor introduced me to his wife, daughter, and her husband. The family of four had come to the US together, and the younger couple was there allegedly to help the doctor set up the business. In the following six weeks or so, I spent a good deal of time with the young couple and less time with the old gentleman. I did spend enough time with him that we became friends and amazingly comfortable together. At 51 years old, I was right between the older and younger couples.

As I worked with the young couple, I started noticing little things that set off alarm bells. Nothing significant, but I became convinced they both spoke a lot more English than they let on. Also, they steered conversations to my beliefs and background enough that I was uncomfortable. Additionally, they were also clearly anti-Semitic but working hard to be well connected with Jewish World Relief. I became suspicious enough that I decided I needed to talk to the FBI about it, just to clear my conscience, if nothing else. However, I was also concerned that if they really were up to no good, and the FBI already had them under surveillance, they just might think I was part of their operation, whatever it was. So, I also wanted to be sure the FBI had a record of me voicing my concerns if they were spies, just to be safe and secure. It is called "CYA" in the military. Cover Your A....!

So, after thinking about it for something more than a week, and just to put my mind at ease, I picked up the phone and called the Dallas office of the FBI, feeling just a bit foolish. After I was connected to an FBI agent, I went over my background, including the fact that I had counterintelligence training some years back. I also presented the facts that I believed he might need to know. I did not want the FBI thinking they had yet another "loony" on their hands, but I had nothing concrete to present, just suspicions.

"My name is George Guy Thomas. I am a retired naval officer. I was a career cryptologist and worked in signals intelligence. I am not a human intelligence officer, but we did get counterintelligence training. For about the last two months, I have been working with four Russians, a couple in their 70s and a couple about 30, who present themselves as a couple. I am suspicious that the younger couple are here as agents and the older couple are here just as part of their cover. Nothing concrete, but their story just does not jell with me the more I get to know them."

I also told him that if the FBI did have these folks during the investigation, I wanted it very clear I was not part of whatever their game was. I was a proud American, and if they did need someone on the inside, I was already there. I also explained that I wanted it made very clear in any records that I came to the FBI with my suspicions. I explained I hoped to get a job back in the classified world and did not want the fact that I was openly associating with people who subsequently turned out to be Russian spies, if that became the case, to be a big question mark in my file.

We talked for about 10 minutes, and I outlined what I knew about them. The FBI agent agreed my suspicions seemed well-founded, but he was not in a rush to ask me more questions. He was either an outstanding actor (which may well have been the case), or there was no open investigation underway that matched my Russian family. In the end, he politely thanked me for contacting the FBI. He gave me his direct phone number saying if I got anything more specific to give him a call. It was all very professional and probably routine for him. I think we both thought it was highly likely we would never talk again. We were wrong.

About ten days later, the old doctor asked me to visit him at his home. I now know that he planned his request for a time when everyone else would be out for the afternoon. He apparently did not want anyone around, but I did not realize that until I arrived at his apartment and no one else was there, which was very unusual. However, it was evident that he had something he needed to talk about with just me. I had no idea what it was but thought it was probably something to do with business planning.

Then he looked at me intently and said, "I am sure you have friends that would like to help someone who worked for the KGB to defect." It was a statement, not a question, but I was not comfortable at all in answering his statement, so I asked him what he had on his mind. He went on to say that he knew things that he was sure the FBI and the CIA would want to know, but he needed protection, or he would be killed, along with his family, if the KGB found out he was planning to defect. I gave him a vague answer assuring him his secret was safe with me, that I did know how to keep a secret, and would "Tell no one!" I promised to investigate how we could contact the FBI but would do it very discreetly not to compromise him, and I would get back to him very soon.

As one can imagine, the second time I phoned the FBI, the reception was equally professional but much different. It went something like: "About my suspicions that I spoke to you about ten days or so ago. I was suspicious of the wrong folks. It is the other pair, the ones I thought were along just for the cover, who are the real subjects. One of them has come to me and identified himself as working for a foreign government. He wants to get asylum and protection. We need to talk."

"Do you think he is genuine?"

"Absolutely."

I did not reveal the identity of the man who wanted to defect, nor what organization he claimed to be working for, and the agent did not ask me to. Both of us were not comfortable discussing what was obviously going to be a classified subject over the phone.

We met for lunch in downtown Dallas at a place he assured me was secure. He asked me a series of questions about my new friends, many of which I could not provide definitive answers to. I just did not know what he needed to know. He also asked about my background and for my social security number.

He made an appointment to meet with me in my home, which initially surprised me. I later realized the agent wanted to see exactly who I was to better judge my credibility and motivation. What better way than to visit someone in their home? I also suspected that he judged my house on a lake in a gated community that he knew was much less likely to be under surveillance.

After we met again and he grilled me again on everything I knew about the subject, he agreed that the FBI should interview my acquaintance. For just such eventualities, the FBI has linguists, but there were no Russian speakers assigned to the Dallas office at that time. One would need to be flown in from DC.

A day or two later, the agent phoned me back and let me know they had checked me out. Also, there were no Russian linguists available to come to Dallas anytime soon. Apparently, they did not think it was a high enough priority at that time. He went on to say that I could opt out, but if I continued as he wanted me to do, I was automatically back under my non-disclosure oath. Would I be willing to be their interpreter?

He pointed out I already had the subject's trust and knew a

fair bit about him, which could be especially useful during the interviews. I readily agreed as he said he knew I would after looking at my record. Then he asked me to get the old doctor to come to my house alone. I was reluctant to call him as I was concerned his lines might just be tapped by his Russian masters, so I visited the doctor at his apartment. I told him things were moving the way he wanted them to, and I had someone he needed to talk to. We set an appointment for several days later, for a time I knew everyone in his house would be busy.

The following nine weeks were very interesting, to say the least. We held a series of meetings at my home with the agent coming in with more and more specific questions about how much the old gentleman knew about the KGB and his field of expertise. I also reported everywhere I went with any of the four and all questions they asked me. The FBI wanted to know what this potential informant's value was before it wasted any more resources.

To this day, I do not know if the agent was playing a role or was genuinely skeptical. Still, the old gentleman was basically told his information was of no particular importance. He needed to be more forthcoming if he was going to be provided asylum. Finally, after obviously a great deal of soul searching on the old gentleman's part, he came up with the level information the agent was looking for and was, I think, sure was there.

"You must not tell anyone, even my family, but the reason the KGB let me come here is that several of my previous students that I recruited to the KGB while teaching in Pakistan, Afghanistan, and Iran are now here, in the US, working to establish both Hamas and Hezbollah support organizations. I am here to re-establish contact with them and bring them back into the employ of the KGB."

The agent turned to me and asked, "Are you sure that is what he said?"

I was more than a bit taken aback myself but managed to

say, "Yes, I am sure that is what he said." In fact, I had asked him to repeat himself because I wanted to be sure I understood what he said.

"I think we have enough information to determine his usefulness. I suspect the Bureau will be sending specialists in this area to talk to him, including people who are fluent in one or more languages in which he is fluent, so we will not need to meet again. Thank you for your service to our country. We will contact you if we need anything else."

And that was that.

I did need to contact the agent about a year later when I went back to work in a job that required a high security clearance. The Defense Investigative Service (DIS) agent doing the background investigation on me kept coming up with references from several of my friends in Dallas that I had recently worked very closely with some Russian nationals setting up a company. Which, of course, was true. The DIS agent rightly wanted to know more about this part of my life, but in that it was within a year of when I had last talked to the FBI, I knew I really could not talk about it in any detail at all. I asked the DIS agent to speak to the FBI agent that I had been working for.

A month or so later, I met with the DIS agent again, and he again asked me about my association with the Russians. I asked him if he had contacted the FBI, which he had not. This act repeated itself at least twice more until the DIS agent finally did call the FBI at the phone number I had given him months before at this point. Suddenly there was no more problem with me getting my clearance back.

****

In March 2003, my wife, Clelia, and I were in our den watching the evening news. That evening it was led off by video footage of a Deputy Attorney General of the United States delivering a statement that the FBI had just arrested 23 people in

Richardson, Texas, for spying for Hamas and Hezbollah. He went on to say that the ring had been exposed by a tip from a concerned citizen in late 1993. I let out a whoop of joy, but then decided I really could not tell her why I was so happy. We had only been married for a few years, and she had no idea I had anything ever to do with intelligence work at that time.

When we first met, I had told her I was a system engineer at Johns Hopkins University's Applied Physics Lab (JHU/APL), which was factually correct. However, she is Peruvian and just learning English at that time. Somehow "systems engineer" translated to mean I was a maintenance man working on air conditioning and heating systems to her.

Indeed, I went to work with no tie, so I could not be professional in her mind. So, she was a bit shocked when she learned I was, in reality, a research professional working on classified systems and even more shocked when she learned two years later at a conference in September 2005 that I knew directors of both NSA and CIA, and that they knew me by my first name. This occurred when Admiral Bobby Inman, the former Director of NSA, and Deputy Director of the CIA, pointed me out in the crowd and said some highly complementary things about me.

Clelia had no idea who Admiral Inman was and turned to me and said, "He really likes you, Guy! Where did you work with him?"

I said, "The last time I worked directly with him, he was director of NSA. I was basically a behind-the-scenes go-between him and Admiral Turner.

"Who was Admiral Turner?"

"He was Director, CIA."

She was totally unprepared for that answer. She may not have fully understood what NSA is at that time, but the whole world knows what the CIA is.

"You know a Director of the CIA, and he knows you by your first name? " The look on her face was priceless!

I could have knocked her over with a feather! I had been "in the shadows" of the intelligence world for so long it had never occurred to me to tell her anything about my past other than I served in the Navy on ships, submarines, and airplanes, and that I had also served with the Air Force. I did tell her a few "sea stories" about my time in the Navy and Air Force.

However, it never even crossed my mind to tell her anything about my background in intelligence. I had not actively hidden it from her; it just never occurred to me to tell her anything about it, just as I never discussed my duties with Linda, my first wife.

In those days, it was explicitly forbidden, but at least Linda was aware of most of the schools and training I had gone through, so I am sure she had a good idea of what I did. But, once in the shadows, you stayed in the shadows in those days.

Introduction to the post 9/11 portion of my life

The first 14 chapters of this memoir dealt with my life in the classified world, but this last chapter will address the birth of satellite AIS (S-AIS) and will touch on the subsequent creation of the "Collaboration in Space for International Global Maritime Awareness (C-SIGMA) concept, the two big parts of my life in the unclassified world post 9/11.

From 12 September 2001 (the day after 9/11) until 22 April 2005 (the first C-SIGMA meeting), I concentrated on the many maritime safety and security issues related to counterterrorism, counter-piracy, and smuggling. I was introduced to two other concerns, the need for environmental and resource protection and conservation, by Norwegians at the first C-SIGMA meeting. Until then, I had not thought about the importance of protecting and conserving the maritime environment and its resources. Indeed, I knew almost nothing about those closely related but different subjects. However, by the end of 2005, I knew these two issues to be huge, fundamental problems.

I also realized a single global maritime awareness system could address all four concerns, and S-AIS would provide the geolocation and identification data that would be the crucial underpinning of this system. Thus, since early 2006, I have focused on building that structure to address all four of these problems. I call it C-SIGMA. By then it was then clear that the rapidly expanding number and capabilities of Earth observation space systems, with S-AIS providing the crucial component, were key to making the maritime world much more transparent. And marine transparency was a huge part of solving all four of these problems. You can make all the laws you want, but if there is no way to enforce them because you cannot promptly detect the rules being broken, you are probably wasting your time. S-AIS is a significant component of knowing who the lawbreakers are and where they were at specific times. Crucial information to law enforcement both in aiding apprehension and as evidence in courts of law.

All these efforts are very much in play on the world stage today. Indeed, the expansion of Earth observation space systems is expanding at a fantastic pace. I plan to finish a second book on just these subjects to record the history of those subjects. I also want the second book to foster a focused discussion on the Earth observation space system's increasing utility to improve safety and security in the global maritime domain dramatically.

These benefits are already accruing now in many places, especially in Europe. The European Maritime Safety Agency (EMSA) and FRONTEX, the European equivalent of a combination of the US's Coast Guard and Customs & Border Protection (CBP), are leading the way in developing better ways to employ unclassified space systems to enhance safety and security in the maritime domain.

Another organization that is doing an excellent job of using space systems to this end is Global Fishing Watch. Indeed, now that they have branched out further than their original mission of protecting the world's fish resources, to include protection of the maritime environment and its resources beyond just fish, they are coming remarkably close to what C-SIGMA initially proposed in 2006.

NATO ACT and its Combined Joint Operations from the Sea Center of Excellence (CJOS COE) has also set out to expand its Maritime Security Regimes effort to look at how to increase the coordination of the maritime safety and security efforts of all nations of goodwill, including the countries of Africa, South America, as well as those of the Indo-Pacific region such as India and Japan. The United States National Maritime Intelligence Integration Office (NMIO) and several of NATO's maritime-centered Centers of Excellence and its Centre for Maritime Research and Experimentation (CMRE) have joined this effort.

This effort, focused on increasing data sharing across the maritime world, now has a subset looking at bringing in space-derived data. This is a good start. The above efforts will only increase both the rate of the development of these capabilities and the availability of them to all who use or benefit from the world's waterways. In the long view, that means nearly everyone on Earth. This is a worthy goal, indeed.

For an excellent examination of why maritime security is still an issue of global importance in the 21st century, I strongly recommend "The Outlaw Ocean" by Ian Urbina, published in 2019. The title is a perfect description of the author's thesis. ***The oceans are lawless and need help***, which he does an excellent job of illuminating via one sordid episode after another.

The Birth of Satellite AIS (S-AIS)
Chapter Fifteen

Satellite AIS (S-AIS) and how it came to be is the focus of this final chapter. However, I believe it is both valuable and necessary to understand the historical context of when, where, why, and how I conceived and initiated S-AIS, so bear with me for a few pages.

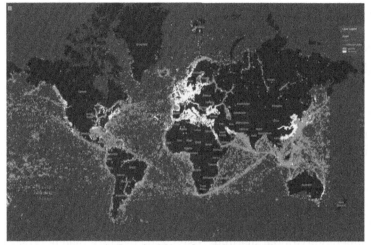

A Day's Worth of S-AIS Tracks

Many other things have changed after 9/11, and my life was undoubtedly one of them. The morning after that terrible day, 12 September 2001, I arrived at my office at the Naval War College's (NWC) Center for Naval Warfare Studies (CNWS) wondering if my life, and that of my many friends in or with the military, were going to change much because of the horrific terrorist attack the day before. I suspected the change would be significant, but I had no idea how substantial it would be.

Dr. Dan Smyth, the Center for Naval Analysis (CNA) representative to the NWC and I, the Johns Hopkins Applied Physics Lab's (JHU/APL) representative, shared a converted classroom as an office in CNWS with three other men who worked as a senior analysts/subject matter expert in a range of fields: anti-submarine warfare, surface warfare, anti-air warfare, etc. A senior NCIS agent was also there with us.

Dan was usually there at 0630, well before any of us. However, on the day after 9/11, he was not in the office when I got there at 0745, and I wondered why. He and I had left the building together on 9/11. We were all sent home early to look after our families on that dreadful day. Security considered the NWC a high-value target and none knew what would happen next that terrible afternoon.

As we walked out of the building and headed home, Dan and I agreed that the US's maritime assets would make exceedingly high value/high payoff targets for terrorism. His first words when he finally walked into our office at about 0805 that morning were: "Do you remember what we talked about as we left yesterday?" Without waiting for a reply, he went on: "Well, the same thought occurred to the President, and yesterday afternoon he tasked the CNO (the Chief of Naval Operations) to conduct a study into the vulnerability of our maritime assets and develop a plan to protect them." He continued: "The CNO has tasked the President (of the NWC), and late yesterday he called me at home to task me to lead the response to President Bush. I have just come from the NWC president's office. We are to immediately organize three war games to define the problem and develop a strawman concept of operations. Guy, with your wargaming and surveillance systems experience, I will need you to be my principal deputy." Then he paused to take a breath. He was excited, and upon hearing his words, so was I.

It took me a few minutes to get my head around the fact that Dr. Dan Smyth, my officemate, had just been officially tasked via the Navy's chain of command to respond to President Bush's order. And that order had arrived via the CNO, Admiral (ADM) Vern Clark. We were to identify the nation's vulnerabilities to maritime terrorism and develop a plan to counter/mitigate them, and that I had just been given a critical role in the response. Wow! Talk about "seats on the front row!" I was delighted! It seemed my professional life had been preparing me for this moment.

The CNO recognized that to do a vulnerability analysis of the maritime assets of the United States, we were going to need input from everyone with a stake in the US's maritime world. These stakeholders included governmental agencies from the federal down to the individual city level and many private entities such as shipyards, port, and harbor managers, shipbuilders, brokers, suppliers, etc., both private and governmental, civilian, and military. He knew the Naval War College held both wargames and meetings involving many of those organizations and decided that the best way to approach this problem would be via a series of wargames attended by all the "stakeholders."

Thus, the Chief of Naval Operations (CNO), Admiral Vern Clark, assigned the Naval War College to respond to President Bush's tasking. The CNO instructed the NWC to use its War Gaming Department and all elements of the Center for Naval Warfare Studies, the NWC's research arm, to answer the order from the President. I was on loan to CNWS from Johns Hopkins' APL. Thus, I was immediately directly involved.

My section was to work in close conjunction with the War Gaming Department, but we were "Lead." And I was lead on developing the required wargame program. Dan and I set to organizing the first game, which was 12 days away. I immediately headed across the street to McCartney Little Hall, the War Gaming Center's site, to start designing the game with its staff of professionals.

Dan got on the phone to gather names on who should attend and start sending out invitations. He had the more formidable job.

The list of stakeholders in the maritime world is exceptionally long, and he tried to invite senior representatives from all of them.

The NWC's War Gaming Center was established in the 1880s and is renowned as a significant war game innovator. It is well versed in holding complicated wargames with a range of classification levels simultaneously, from entirely unclassified to highly compartmented intelligence. Indeed, in 1982 I had created a pair of cells at different compartmented intelligence levels. One cell inserted an accurate portrayal of special operations forces' capabilities into war games as required: the other added all pertinent aspects of satellite capabilities at the right level of technology for the year of the game (today, next year, five years from now, ten years, etc.) into war games.

By the time I left the NWC in April 1986, I had almost five years of experience with NWC wargaming. By 2001 I had eight additional years of wargaming and follow-on analysis experience. Venues included two years at the Joint Electronic Warfare Center (JEWC), reporting directly to the Joint Chiefs of Staff (JCS) organization, and then one year at the Naval Air Research Center at Patuxent River. Finally, and most importantly, I had five-plus years at the Warfare Analysis Lab (WAL) of JHU/APL, the acknowledged world leader in multi-facetted wargames stretching from pure research through applied research, to tactics, operations, strategy, and policy.

In the next several months my wargaming experience would be expanded even further. We designed the war games conducted immediately after 9/11 to include all US maritime domain elements, plus those of our closely associated allies (Canada, England, etc.). All stakeholders in both the public and private sectors, including all branches of the US, state and local governments, and all associated private parties such as ship, shipyards, port and harbor owners, builders, and operators, were invited to participate. They were tasked to help us identify all known vulnerabilities and discover unknown ones.

These next several months were a blur. Designing this war game series was "a walk in the park." But their execution, the capture of the pertinent data, and the analysis to make sense of the data we had collected in the games was tedious, if not demanding. We knew we needed to learn a great deal from our participants in these games, and we mined the data generated by the wargames and symposiums very diligently.

The CNO had selected the NWC because of its ability to hold multi-classification level war games for large groups, and we were expecting well over 1,000 people. Over the years, this capability had been developed and regularly exercised in the Global War Game series held each summer since 1979. I had participated in many of these games as a surveillance systems and electronic warfare subject matter expert (SME). I was also the designer and game director of a series of Command, Control and Communications Counter Measures (C3CM) games from 1983 to 1986 that were directly pertinent to the post 9/11 games.

For those games, I developed methodologies to input simulations of both Space Systems and Special Operations play as accurately as possible. Both of those capabilities have highly classified components. However, for this information to be included on the main game floor, it had to be distilled down to the Secret level. I spent significant time developing methodologies to accomplish this objective while maintaining the security guidelines. This experience proved to be very useful as we set up the games in the Fall of 2001, fifteen years later.

Maritime infrastructure is both exceedingly vulnerable and very significant to both the US and the world's economic health and well-being. Dr. Dan Smyth was in charge of the overall effort to draft the response back to President Bush. He tasked me to co-lead the organization of the post 9/11 games and act as a subject matter expert on surveillance and communications systems, my core experience for over 30 years at that point.

Many wargames are designed and held to train and inform

the participants themselves. Organizations also have wargames for research purposes. Research wargames are often held in series. The series allows the participants to discover what they do not know about a subject and then apply those lessons to the following game. This process is repeated until a sufficient understanding of the issue is reached.

In that this was a research wargame to develop understanding and insights, one primary task was coordinating the dozen or so NWC personnel assigned to take notes, the rapporteurs. They are critical to any wargame success from which you want to extract lessons for follow-on actions. My job was to be sure they knew how to be a rapporteur before the games started. Then, take their notes after the event and boil them down to the lessons we needed to understand. (Knowing and Understanding are two entirely different, but related things. Understanding flows from Knowing.!)

The first part of this task, designing the game and organizing the rapporteurs, was easy. The NWC personnel included many who had been rapporteurs before and were absolute pros at it. The second task, to refine the inputs into a new concept of operations for the nation, was the real challenge. As the lead rapporteur, the principal recorder, I worked with about a dozen others to capture the salient points and counterpoints generated during the war games. This last task is an essential function in any wargame, especially so for this one as it was from these notes that we would build the concept of operations to construct the President's answer. As the planning for the games progressed, I realized I would be right in the middle of something incredibly significant, not just to the Navy and the nation but also to the world. I was delighted to be doing something so valuable. On the day of 9/11, we all felt useless, and now we had a focus. It was a great feeling!

It usually takes three to four months to plan and execute a significant wargame, but we had 12 days, and it was going to be a huge one, with many diverse elements. We all "fell to" as said in the Royal Navy and were ready by 9/24. It was a very hectic war game because many participants had never played in one. However, the wargaming staff had experienced this level of inexperience and knew how to deal with it via training. They held instructional briefs at the onset and then individual coaching during the first day of gameplay. Thus, we were able to bring everyone into action by the end of the first day.

The most surprising fact to most of the attendees that came out of the first wargame (24-26 September 2001) was the explicit knowledge that no agency tracked and positively identified ships until they were basically inside our harbors. We recognized we needed to have more advanced warning, and we generated a requirement to have all ships identify themselves 96 hours before entering port. The US Coast Guard very quickly published a "Notice to Mariners" establishing that rule.

The requirement to have ships identify themselves via that system when they were at least 96 hours before arriving in the port and then continue broadcasting their position every X many hours (3? 4? 6?) was the one generated concrete requirement in the first wargame. Taking the input from many people, I drafted this new requirement, and the attendees approved it. Interestingly, during those discussions, it appeared to me that I was the only one present who knew that the INMARSAT satellite communications system had a built-in position location reporting system that the ship could set to report automatically.

We decided to ask the International Maritime Organization (IMO), an agency of the United Nations, to implement this rule as soon as possible. The US Coast Guard took this requirement for action but warned that it might take some time as the IMO was not known for moving quickly. We all thought that the IMO would act much more swiftly in these tense times than they subsequently did. The IMO eventually enacted the requirement, but it took several years, not months! Its official IMO name is Long Range Identification and Tracking (LRIT). Some people see it as a competitor to S-AIS, but it is not. It is fundamentally different. LRIT requires a conscious action by the ship.

It needs to consciously set the satellite communications system to report its position every four hours. We designed S-AIS to pick up the AIS beacon that all vessels over 300 tons engaged in international commerce are required to have on anytime they are approaching within 2,000 NM of shore. Indeed, most ships leave it on as an extra, automated lookout. AIS broadcasts its location and ship's identity at least every 10 seconds if it is underway.

We held our second war game on 1-3 October. There were two primary results. At the end of the second war game, I believed we had enough material to commence writing a draft concept of operations and told Dan so. He concurred and set me to drafting it. He was wholly tied up with the administrative requirements, which are substantial for a game this size. He was happy I was willing to take on this task. It was THE core task.

One of his primary duties was to ensure the right people were attending and were assigned to the most useful roles, given their experience. It is much more complicated than it sounds, what with egos, busy schedules of senior people, and all. You also have limited knowledge of the players' skills and personalities, especially in this series, so much of this is skillful guesswork.

It was one of the most diverse groups of participants ever assembled, although the Global War Games of the 1980s, run each summer there at Newport, were almost as varied. However, the Global games had the advantage of having most of its participants familiar with wargaming and its analysis aspects. This group did not. For this series, we had many civilians with no real idea of how to define our problem set, much less how to address it for analysis.

As we processed through the first two wargames, my team of several assistants and I tried to keep a combined, consistent record of the leading emerging points. I immediately started organizing them into a rough concept of operations. One fact stood out to me. We needed a better way to track all ships approaching our shores. During Game #1, Don Cundy, the electronic systems department head at the USCG Research & Development Center, mentioned a new line-of-sight ship tool called the Automatic Identification System (AIS). The United Nations' International Maritime Office (IMO) had recently mandated it as a collision avoidance and ship traffic control tool. The USCG was making plans to use it for harbor control and improve their situational awareness of what was happening in and near our ports and harbors. I asked him to present a brief on AIS at Game #2 on 1 to 3 October.

At that brief, on the last day, we learned the IMO had directed AIS to be installed on all ships of the "Safety of Life at Sea" (SOLAS) class by February 2004. SOLAS class ships are defined as all commercial ships displacing over 300 tons, all vessels carrying six or more passengers, and all tugs over 600 shaft horsepower. Vessels displacing 300 tons are close to 100 feet in length, so most smaller vessels were not included. However, many individual nations have since modified downward the carriage requirements. As of 2018, the US carriage requirements are now down to all vessels over 65 feet in length.

As the USCG R & D Center presented its brief, I thought back to my time as a space systems subspecialist in the Navy specializing in signals intercept and analysis. It instantly occurred to me that we just might have the answer to our offshore tracking problem, and I asked him if I could get an in-depth brief on AIS. He offered to give me one the next day at 0930 at his office in Groton, CT, about 60 miles from the NWC, and I readily accepted.

Early the following day, I drove down to Groton. The appointment was for 0930, but I arrived ninety minutes early. I was concerned about the 8 to 9 AM rush hour traffic on Interstate 95. As the Officer in Charge of a very busy detachment in Atsugi, Japan, I knew arriving early could be very disruptive and might even be seen as rude. This was especially true that day as I knew he had been with me in Newport for the previous several days and might like to get a bit of time in his office to catch up. I know I always did after a day or more away from my office. So, I just read in my car. I had brought a book along for that eventually.

I rang the bell right at 0930, and they were ready for me. Don had Dave Pietraszewski, the US's AIS expert, give me a brief and then a live demonstration of the signal. Dave displayed the AIS signal on a radar scope and then output its data stream to a printer. He also demonstrated how one could click on the AIS icon on the radar scope, and all related data would appear on the screen, not unlike air contacts with their Identification-Friend- or-Foe (IFF) system.

When the briefs were over, and they asked one last time if I had any questions, I asked the question that had been running around in my head for the past 24 hours. "Anyone ever thought about putting an AIS receiver on a low-Earth-orbiting satellite?" The answer from around the room full of engineers was that the thought had occurred and been dismissed. There would be too many signals of the same frequency and power arriving at the satellite's antenna simultaneously, and all the satellite's receiver would hear would be "white noise." The receiver would be totally overloaded.

From my time at NSA and especially my time working on upgrading the communications system on Air Force One, I was sure I both knew how the signal could be isolated and decoded and the guys that could do it. However, I did not argue with the gentlemen in Groton for two fundamental reasons. I was not 100% sure I was right, and I was not sure of the classification of what I knew. It may well have been Top Secret or at the very least company proprietary to E- Systems.

Besides, I was their guest, and that would have been not polite, so I just said I thought there might be a way around it and was going to investigate it a bit more. They skeptically wished me luck, and I departed for the 60-mile drive back to my office at the

NWC.

Two other things pertain here: First, I had known since at least 1968 that ships did not have an identification beacon on them like aircraft with their "Identification Friend or Foe: (IFF) system. Second, about ten years before, I had helped develop the plans for putting the first multi-band phased-array communications antenna ever installed on an airplane. It was for Air Force One, the President's plane, at E-Systems in Greenville, Texas. I had also sat in on many discussions with its partner Mitsubishi Electronics (MELCO) in several places in Japan. Those discussions had included several on how that early phased array antenna would gain sufficient signal isolation from the signal noise floor. I thought that process just might be applied here.

I hurried back to my office at the NWC and, even before I sat down, placed a call to Neil Cooper, one of the chief systems engineers at E-Systems. E-Systems was the developer of the communications systems on AF ONE from the beginning, back when AF-ONE was a Lockheed Super Constellation. Neil was as competent a man technically as I have ever met. I explained what I was thinking and what I now knew about AIS and asked his opinion. His answer was basically, "Physics is physics! It makes sense to me! Go for it!" So, I did.

I first used Google to look for an active satellite with a VHF system. There was only one, no matter how I asked the question. It was ORBCOMM, headquartered near Dulles Airport in Virginia. I immediately phoned ORBCOMM, but the first three or four people I talked to there at ORBCOMM wanted me to sign a non-disclosure agreement before they "would give me the time of day."

After those three or four unsuccessful attempts to talk with various people in their engineering department, it dawned on me. I was talking to the wrong people. I phoned ORBCOMM back for the 5th time and asked if they had a person in charge of business development.

"Yes, Sir, we do. His name is Greg Flessate."

"Could you put me through to him?"

The operator immediately put me through to Greg, Director of business development for ORBCOMM. This time I did not immediately start talking about technology. I started with, "Greg, I have an idea that could, conceivably, make ORBCOMM a lot of money! Are you interested?" Greg's response was just what I had hoped it would be. "You have my attention. Tell me more."

After about a 20-minute chat, he invited me to visit ORBCOMM, near Dulles International, as soon as possible, and I arrived there less than a week later. At that meeting, he agreed to place an AIS receiver as a "ride-along package" on their next satellite launch to replenish their M2M/IoT constellation if I could find the money to design and build the receiver and cover the installation and integration costs. We were talking about roughly $10 million. The time-window for getting this done was about 24 months. (Thankfully, over the next two years, that time window slipped about another two years.) Over the next 27 months, I briefed everyone I knew from my time with the Fleet Battle Experiments and the several Maritime Terrorism war games and seminars of late 2001/early 2002, asking for money to build my satellite to no avail. There were many hundreds of people. I was getting discouraged.

In November 2001, the USCG also asked me to help the Naval Underwater Warfare Center) there at Newport, part of Naval Sea Systems Command (NavSea), to help them install an AIS transceiver in a submarine preparing to depart for a patrol, and I did. We got them a specially modified transceiver with a 3-position switch, "On-Off-Receive only." The captain's post-patrol report was all we hoped it would be. "Don't leave home without it!"

Over the next 27 months, I briefed everyone I knew from my

time with the Fleet Battle Experiments and the several Maritime Terrorism war games and seminars of late 2001/early 2002, asking for money to build my satellite to no avail. During that period I briefed over a thousand people. I was getting discouraged.

During that period, I also briefed anyone and everyone I could find in the USN and USCG on both the utility of AIS and the need for Satellite AIS (S-AIS). I often mentioned the submarine test and got much feedback, a large part of it quite negative. Almost no one had ever heard of AIS, and the idea of putting an AIS receiver in Space sounded like craziness to most people.

I distinctly remember saying many times that it would be an excellent idea to have AIS "full-on," send and receive, in congested traffic areas where a warship's location and probable identification were already compromised. I also distinctly remember explicitly mentioning the mouth of Tokyo Bay, which I had transited in two cruisers and twice in submarines, and repeatedly flown over in four different USN and USAF reconnaissance aircraft in my career. It was the most congested waterway I ever saw. However, many people have told me that other narrow waterways, such as the Straits of Malacca, the English Channel, and Straits of Gibraltar, are even worse.

Nearly the only encouragement I received was from my home organization, Johns Hopkins University's Applied Physics Lab (APL). I was required to return once a month to discuss the issues of the day and receive guidance from APL as to what information they desired from me. My command chain was Chris Latimer, Steve Biemer, and Russ Gingras, the man who had hired me. He was my department head. I regularly conferred with them on the progress of my work at Newport and on finding funding for my "wild idea." They were all encouraging, urging me to keep trying as the prize was worth the effort.

They also referred me to specific APL members who had

experience with space technology to verify what I was considering was technologically feasible. I also received excellent graphics art support from APL's graphics art department. The people there are true artists and knew how to sell a project with their skills. It was their work that I was dragging all over the US looking for money.

The other encouragement I received came from APL's technology research rival, the Naval Research Lab (NRL). I had briefed Mr. Pete Wilhelm, the Director of the Navy Center for Space Technology. During my time on the Naval War College staff fifteen years previously, I had interacted with him, but I had not thought him out. However, when he came up to the NWC for a conference about six weeks after I had my S-AIS idea, I seized the opportunity and briefed him. He liked the concept and put me in touch with Andy Fox, one of his principal technical assistants. Over the next 30 months, we often conferred as to where I might find the money. They even offered to assist ORBCOMM in designing their satellite, but that never came to pass, for reasons I never understood at all. Jealousy? Fear of compromising company privileged information? A combination of both of those? I never knew for sure.

By mid-November 2003, I had just about given up hope of ever finding money to build and install an AIS receiver in a satellite. Then I struck up a conversation with CAPT Tom Rice, USCG (Ret.), a new colleague in the newly created Maritime Domain Awareness Program Integration Office (MDA/PIO), an organization enroute to being a joint/interagency/ inter-departmental office hosted by the USCG to bring order out of the mishmash of maritime awareness programs across the US government.

I had taken early retirement from APL, taking a significant cut in salary to come to work there a few weeks earlier, but this was one of the first times I had had time to sit and talk one-on-one with Tom. He had recently retired as the head of maritime Navigation Systems for the USCG (and thus for the US). As we sat discussing the task before us, he offhandedly remarked: "If we could just find a way to extend the range of AIS, our ship tracking task would become a lot easier." My response was entirely predictable if you knew my history for the past two-plus years.

"Are you trying to upset me?"

"No, why would you ask that?"

"Because I bet I have briefed at least 50 people in the USCG, including your deputy, about my idea of putting an AIS receiver on a Low Earth Orbit (LEO) satellite which solves that exact problem."

"Would that work? What about co-channel interference?"

"I know people that are a lot smarter than me about this, and they assure me the signal can be isolated and read."

"How come you have not told anyone about this?"

"I bet I have briefed over 1,000 people in the last 26 months, involving every branch of the Navy and USCG, including several folks who worked for you."

"No one told me anything. I would like to see your brief myself."

"Don't have it here, but have a CD sitting on my desk at home. I can bring it in tomorrow."

"Great, See you at 0730 tomorrow. Right here."

The following morning Tom Rice's immediate reaction to the brief was most positive. His remark went right to the heart of the matter.

"You need to brief the right guy."

"OK, smart guy. Who is the right guy?

"Our new boss, Jeff High. The man we were talking to

yesterday."

"You obviously know Jeff High very well. How about setting up a time for me to brief him."

"Be right back,"... and he disappeared for seven minutes.

Walking back into his office, his words were short and to the point:

"0745 tomorrow. Be ready!"

"I am ready now!"

"Jeff is a very busy man. He will see you tomorrow."

The next morning, I briefed Jeff High right on schedule. He stopped me just less than halfway through my satellite AIS brief with that all-important question:

"How much would this cost?"

"I have discussed this in detail with ORBCOMM, and as a "ride-along" package with their planned constellation replenishment mission, it would cost less than $10 million to design, build, launch and operate the S-AIS package for five years."

"That is very reasonable if this satellite performs as you describe."

"Sir, that is what I have been telling folks for the past 27 months. It is a terrific deal, with an extremely high potential payoff."

"I am drawing a circle around $10 million in my budget right now. Get with Legal and Acquisition and come back with a firm number. Try not to spend it all! If you are right, this may well be the best $10 million the US Coast Guard has ever spent."

After 27 months of searching, I finally had the break I needed. Dealing with lawyers and acquisition personnel is not my idea of fun, but they were very professional in assisting this "newbie" (me), and by the end of March 2004, we had a signed contract.

At my suggestion, we did take one additional intermediate step I believed very necessary. I asked for permission to fund a feasibility study at JHU/APL to ensure there were no unforeseen problems. I had already discussed this with Steve Biemer at JHU/APL and had a cost figure to give the USCG on the tip of my tongue. $110,000. I also requested permission to bring ORBCOMM into the study as they were the most experienced company in the US, and probably in the world, on VHF satellites. In that, I had already written a "Sole Source Justification" for ORBCOMM, the Legal and Acquisition folks agreed, and we set up the team.

ORBCOMM sent their best, David Schoen, the leading engineer behind all three for the low earth orbit (LEO) communications satellite systems then in existence. He had rotated through employment at Iridium and Global Star as well as ORBCOMM, developing the signal processing math for each satellite constellation in turn. He went on to be the Chief Technical Officer for INMARSAT, a major satellite communications company.

APL reached into their deep pool of talent and assigned three young and brilliant PhDs. I do not think any of them were 35 yet. Maybe not even 30. We had our answer after less than six weeks of study. No one could find any reason why S-AIS would not work. APL developed a formal brief, and I briefed Jeff High with it. He signed the contract that I had worked out with the USCG and ORBCOMM lawyers. I wrote over 90% of the contract using examples from other similar agreements and then submitting my work for review by my USCG Legal and Acquisition advisors. It is the only contract I have ever written. The rest is history, but even then, it was not all smooth sailing.

Just after we signed a contract with ORBCOMM to design,

build, launch, and operate the first Satellite AIS system, the US Joint Spectrum Centre (JSC) wrote a technical paper highly critical of our APL paper saying that S-AIS would never work.

It was quite a shock. I asked my APL team of very bright young PhDs to do a total review and comparison of our analysis, which said it was feasible, with theirs, which said it was not. I was very anxious that I might have just wasted 6.4 million dollars. If so, my career in the science and technology (S&T) field was over with a thunderclap.

I should not have been so concerned. I had a great team working on the problem. But it did take that team three days to find the difference in the two analyses. The JSC had misplaced a decimal point in the definition of the signal density at that frequency. They had it ten times denser than the acknowledged value. Our calculations were correct; theirs was in error.... If you moved the misplaced decimal in their algorithm to the proper place, the two studies agreed! I went to sleep for the first time in three days. Still, this incident made me very anxious to see actual data from the satellite.

There was another existential threat to the S-AIS program. CG-2, Coast Guard intelligence, had just been formally admitted to the "Intelligence Community" (IC), and they had minimal background in sophisticated signals intelligence collection and analysis. Worst yet, they did not know what they did not know. I had the clearances to visit their spaces, and I did try to discuss how we could work together. But they did not want to hear any suggestions, no matter how gently I worded my offer. A few old Navy signals intelligence professionals were scattered about USCG HQ, and they all privately agreed with me. However, the guys who had come up through the CG law enforcement pipeline, all good men, were clueless about sophisticated data systems such as AIS, and all the ex-Navy guys all knew it.

The CG-2 folks were now dealing with people in this area at NSA for the first time. These were people that I had known for many years. Some of the senior leaders had even worked with or for me. They reported to me that the USCG guys were daily demonstrating their lack of experience and knowledge. They were not stupid by any means, but they were ignorant. There is a big difference. The NSA folks pointed out to CG2 that they ought to work more closely with me, but that just made matters worse.

CG2 quickly became very jealous of my stature there and told the NSA that I could not visit there without one of them being with me! The NSA guys just laughed out loud at that stupid edict and ignored it. I was in an inter-departmental organization, the MDA/PIO, so I ignored it too, but CG2 did hold my clearance, so it did get a bit dicey. Eventually, I asked Jeff High, my new boss (a CG SES-3), to provide me with cover. In that he was senior to the head of CG2, who did not even know his juniors were restricting my access to NSA, that problem went away.

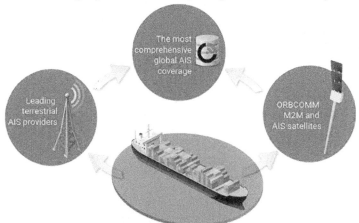

However, mid-level civilians at CG2, working with mid-level civilians at NSA, tried to kill the S-AIS program by getting it declared a SIGINT program, requiring a "Special Compartmented Information" (SCI) clearance to know anything about it. This idea, too, was laughable, but it became a real, direct threat.

The matter eventually blew up, and the issue went all the

way to the Secretary of Defense, and in late 2004, he directed both NSA and me to prepare memos defending our positions. It could be no longer than 3 or 4 pages. I forget precisely how many because my paper was less than two. NSA's took the limit. Mine said that I did not care what NSA did with the AIS signal behind closed doors, but there was a very valid need for it in the open, uncleared world.

Besides, I had been describing how I thought S-AIS could be used in the open market to anyone who would listen for about three years at that point. And I had cleared it with very senior people at NSA in December 2001, before I ever opened my mouth in public. To try to classify it now was very much too late. I pointed out that the effort to classify it ex post facto alone would generate much more interest in AIS, and I did not think NSA wanted that. "The cow was out of the barn. It is of no use to close the barn door now."

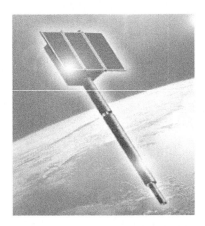

ORBCOMM's first satellite with S-AIS

Members of the OSD staff tell me the Under Secretary of Defense for Intelligence read both memos in a meeting at the Pentagon and immediately agreed with me! I breathed a big sigh of relief, and the plans for ORBCOMM to add AIS receivers to its next satellites went forward.

I understand from Canadian friends that when ComDev, the parent company of exactEarth, the second company to operate a commercial S-AIS constellation, went to get permission from the Canadian government to create their S-AIS constellation, almost the exact same thing happened there in Canada. In both cases, common sense won out.

Today S-AIS is widely used across the maritime world and has, indeed, created the most significant paradigm shift there since radar, the steam engine, and the screw propeller. Yes, even more game-changing than GPS. GPS allows ships to navigate with much better accuracy, but it still left the ocean opaque while S-AIS is making the world's waterways much more transparent. It enables the United States to understand who was approaching its coasts and ports, just as intended. However, it has since become a ubiquitous tool for an ever-increasing array of maritime applications.

It should be noted for history's sake that NRL, with whom I had also been working in conjunction on this project ever since I briefed Pete Wilhelm on my idea in the Fall of 2001, beat ORBCOMM into orbit. They put an AIS receiver on TacSat-2, a proof-of-concept demonstration satellite launched on 16 December 2006. That AIS receiver had a problem with overheating and had to be turned off after 20 minutes of operation on each orbit. But TacSat-2 did prove the AIS signal could be collected from Space. My concern here was whether the NRL used some undisclosed processing to pull the AIS signal out of its very dense signal environment. I tried to get the NRL and ORBCOMM to discuss this but failed.

ORBCOMM originally planned to launch in late 2005. It would have been the first S-AIS receiver in Space, but there were several delays, mostly with their primary payload, not the AIS portion. ORBCOMM was finally ready for a mid-2007 launch of its six "Plane A" replacement satellites with S-AIS receivers via a Russian Cosmos rocket. However, at the last minute, Russia learned that some of the ORBCOMM/Coast Guard team had intelligence agency backgrounds (including me, specifically). The Russians became convinced the ORBCOMM mission was some CIA trick. We offered to give them full access to the downlink, but they were having none of it. They backed the launcher off the pad, dropped it on its side, pulled the six ORBCOMM satellites out of the bus, and put in bags of sand as ballast. They buttoned the launch vehicle back up, rolled it back onto the pad, erected it, and launched it without the ORBCOMM satellites. It took ORBCOMM over a year to get another ride, but they finally launched their six satellites on 19 June 2008.

In the meantime, Space Quest, a small US company based in Fairfax, Virginia, was about ready to launch a test satellite with a programmable receiver in the VHR range, the same range as AIS when they read a story regarding ORBCOMM, S-AIS, and the US Coast Guard. They decided to reprogram their receiver to be able to collect AIS. They successfully launched the first S-AIS receiver, which could operate for an extended period into Space on 17 April 2007, six weeks after first hearing about S-AIS! Indeed, they built and launched two more S-AIS satellites, ArizeSat 3 and 4, 29 July 2009. They are still in the satellite-making business today.

On 28 April 2008, Canada became the first foreign country to launch an AIS receiver into Space with its Naval Tracking Ships (NTS/ CanX-2). Norway was the second foreign country to launch and operate an S-AIS receiver when they collaborated with the Japanese to put their NORAIS receiver on the International Space Station in June 2010, launching AISSat-1 on 12 July 2010 about a month later. All these entities attended our meeting on 22 April 2005 held by the Maritime Domain Awareness Program Integration Office, hosted at US Coast Guard Headquarters. I consider that the date that S-AIS came of age and the concept of C-SIGMA was born. More on C-SIGMA in the next book. It deserves its own book!

SUMMARY

There are now, early in 2022, about 205 S-AIS receivers in Space. The three primary providers of S-AIS coverage are ORBCOMM, exact Earth/Harris, and Spire. Between the three of them, every AIS transmission anywhere on Earth is being collected almost every minute. I believe that the time interval will fall to well below a minute within the next few years.

S-AIS is now routinely paired with imaging space systems such as electro-optical satellites and synthetic aperture radar (SAR) satellites that can image the Earth day and night, rain, or shine. There are not enough SAR sats yet to do this 24/7, but with the increasing number of imaging satellites of all types, SAR, EO, video, plus the new radiofrequency geolocation satellites (RFgeoSats), it is only a matter of a very few years before nearly every place on Earth, including the oceans and other waterways, will be under near-continuous observation/surveillance.

With all three types of these imaging satellites plus the RFgeoSats, there is a natural synergism in the maritime world as S-AIS gives them the added ability to identify the ships imaged. If they are not identifiable, the very lack of an identity, or their attempt to hide their identity via spoofing, can lead to further examination by other Space and terrestrial assets. Those assets can include other space systems, patrol aircraft, and military or law enforcement vessels.

Indeed, attempting to spoof S-AIS is, in many cases, counterproductive for the spoofers as they are calling attention to themselves by the very act of sending out false information. In that many have the mistaken impression that S-AIS is easy to spoof undetected, a few words on that subject are required.

Since the first ORBCOMM satellites with S-AIS as a secondary payload were launched in 2008, there has existed the capability to locate all AIS emitters collected. That geolocation can be compared with the reported position contained within the transmission itself. If it is more than a certain number of miles different, the report is flagged, and that emitter is collected at every opportunity. Generally, once a position is verified to be "true," all following collection of that emitter by that satellite during that collection opportunity is discarded. (A collection opportunity against a transmitter is defined as when the terrestrial transmitter is within the field of view of the satellite). It is the antenna pattern on the satellite which defines its field of view.

The data provided by the AIS signal includes the Maritime Mobile Service Identity (MMSI). The MMSI is a unique 9-digit number that is assigned to each AIS unit. Some of the anti-spoofing software tools contain databases that strive to record the MMSI of every AIS signal detected anywhere, stretching back to early 2001. It is straightforward for these software tools to sort through the history of every MMSI ever broadcast in milliseconds and notice whether there is something amiss. Those transmitters are also flagged for particular attention. False locations and false MMSI generate extraordinary attention in some circles, which cannot be discussed here. To be sure, not everyone has these tools, but they are readily available on the market. I would be delighted to introduce anyone interested in buying their services/tools to several developers. Most of them are friends of mine.

Another separate business area has also arisen. It has many names, but I prefer "Static & Dynamic Data Analysis" (S&DDA). It takes the reports from the S-AIS satellites, from terrestrial AIS, as well as from the myriad static information and data sources in the maritime world such as ship movement reports, brokers reports/records, financial news, and shipping documents and strives to make sense, understanding, and (hopefully) wisdom, from this mass of data. Many of these companies are now using artificial intelligence and machine learning to develop their reports. Some of these systems are now designed to provide their information to their customers in a "tactically useful timeframe." (Long a requirement of mine!)

These many software tools process AIS data for many uses.

Radar and Satellite navigation, even in its most recent form, GPS, are the only other systems that even comes close to the impact of S-AIS in the maritime world in the last 150 years, but while GPS allowed for mariners to navigate with more surety, it left the marine world opaque, while S-AIS is quickly making it much more transparent.

Like satellite navigation, created to improve the US's submarine-launched ballistic missiles' accuracy, S-AIS is rapidly becoming ever more present in the marine world as more and more applications for its data are being conceived and developed. However, while GPS is now found in many applications for all three environments, sea, air, and land, S-AIS's impact is focused on the marine world, causing significant global maritime operations changes.

Indeed, as far as the maritime world's overall operations, the impact of S-AIS in its first dozen years of existence is more significant than GPS at this point in its evolution and has most probably already surpassed it in total. Let us look at just some of the changes: First, it is important to note that while S-AIS was created as a maritime security system, its usage has expanded into many fields.

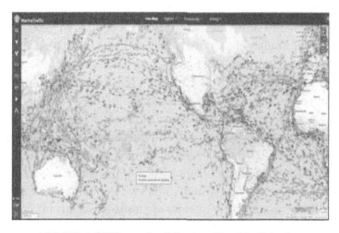

AIS TRACKS on the Marine Traffic Display

Commodities

The tracking of all the world's commodities location and the estimate of their time of arrival at the destination is now down to a very few hours. This detailed information, derived from S-AIS and brokers' records, has allowed commodities traders to be dramatically more accurate in predicting daily prices in many ports of the world. The first few who developed and employed the analysis based on this information were able to reap substantial financial rewards. Now that everyone is doing this, the field is much more level.

Marine Maintenance

Determining when hull and machinery maintenance needs to be scheduled for maximum return on investment because S-AIS allows for very accurate record-keeping of hours of operation at what loading/speed and in what type of marine environment at what average speed (critical to hull maintenance). These calculations are saving the big fleet operators millions of dollars.

Illegal Fishing

Determining when a ship is bound for an area that is environmentally or marine resource-challenged is good to know. That vessel might pose a threat, either intentionally, such as illegal fishing, or unknowingly, such as having traversed an area known to be contaminated with a sea life-threatening biological problem such as a disease, or predatory fish, allows concerned authorities to demand the diversion of the vessel until it can be determined to be a threat or not. Also, fishing boats' patterns of the courses while fishing are very distinct in many instances. Several software tools now automatically recognize these patterns. Global Fishing Watch, the Pew Foundation, CLS, and others have built an array of software tools to aid this effort.

Environmental Protection

Using synthetic aperture radar satellites to detect illegal bilge dumping in controlled waters is now much more effective because the exact identification of the offending vessel and its next port of call can often be determined via S-AIS. When the offending vessel docks, the local authorities can demand to see its bilges and logs. If the bilges are clean, but there is no log entry detailing when and where they have last pumped them, then a citation with a fine is issued often for many thousands of dollars. The Italian Navy reports this has caused a dramatic reduction in the illegal dumping of bilge waste, thereby improving the marine environment in the Mediterranean.

Search and Rescue

S-AIS has also dramatically improved the safety of life at sea by allowing for all ships' locations in an area to be known by all interested parties, thereby permitting the rapid reaction to maritime disasters, large and small. Indeed, AIS transmitters are now being installed on life jackets to assist in the recovery of crew in the water. S-AIS allows for the closest vessels to a maritime problem to be identified and vectored to the site as needed.

Disaster Mitigation and Recovery Support

It is also a powerful tool in disaster recovery operations. In Hatti, S-AIS provided the knowledge of when needed supplies would arrive at the only port in the country still operating but with only minimal capacity due to the damage. S-AIS allowed for the accurate planning of the landing of the most needed supplies first, in priority order. It has been used similarly in many disasters since, including in several countries such as the Philippines, which has been hit several times with hurricanes, and in the Indian Ocean, Japan, and Chile, where tsunamis have severely damaged ports and seacoasts. It is now standard operating procedure in these situations.

Security and Surveillance

Finally, S-AIS is being used worldwide as a primary adjunct

to maritime security operations by allowing for the study of regular "pattern of life" operations to determine when abnormal activity is happening. Thus, surveillance and intercept units can be dispatched with much-improved chances of apprehending wrongdoers, be they smugglers of all types, illegal fishers, illegal immigration, or whatever. This saves wear and tear on equipment and personnel and saves money by limiting the operating time and thus costs of scarce assets. It also raises crew morale because it knows they have a better chance for a productive operation.

Sanctions Compliance

The exact pattern of life at sea that S-AIS allows to be generated is now used to identify international sanctions violators as they need to violate those patterns in hopes of not being detected. Here too, AI and machine learning are being employed to excellent use.

And More.

The list just keeps getting longer, but clearly, S-AIS is making a significant impact on the maritime world.

I must admit, I am immensely proud!

CODA:

S-AIS and machine-to-machine (M2M) satellite communications have recently been combined into a revolutionary package the size of a square beer can. The device is called the MT5000. It was initially known as Hali. With an 18-inch whip antenna, it was designed for the small-sized fishing and pleasure boats of the world. It transmits low-powered AIS to alert larger ships and shore stations within 10 to 12 miles of its location as a safety device. It also broadcasts its position to ORBCOMM's 40 satellite M2M constellation at the same time to allow for the relay of this data to a fleet operator or family members, or anyone else the mariner wishes to keep informed of his/her location. The uses for this new device, the MT5000, have just started to be explored, but I believe it will be another game-changer, especially in the small boat world.

NOTE:

It is interesting to note that both satellite navigation and Satellite AIS were invented at Johns Hopkins University's Applied Physics Lab (JHU/APL). I was an employee there when I conceived AIS, and the encouragement and the assistance that I received from there in many ways as I struggled to sell my idea were invaluable. I might well have quit looking for funds had I not had such a solid backing from Russ Gingrass in my home office and Pete Wilhelm, Director of the Navy Center for Space Technology.

****

I have been part of the global team creating the system I call "Global Maritime Awareness" for almost 20 years. I believe it will be a world champion soon. Indeed, I am also sure that if I were to come back 100 years from now, I would not recognize it just as Orville and Wilbur Wright would not recognize a Boeing 747, an Airbus A-380, or a Lockheed F-35 if they came back today.

Just like them, I had no idea exactly what I was instigating. The Wright brothers and I both saw a need and dedicated our lives to making our dreams happen. It is a great feeling!

The author stands ready to assist in the creation of the Global Maritime Awareness system by any legal organization, any country, or group of nations. The need is there; the capabilities are there. We just need to put them together in an intelligent way.

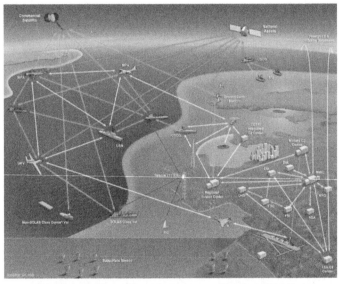

Maritime Domain Awareness (Spring 2002)

Annex 1

A MARITIME TRAFFIC-TRACKING SYSTEM

CORNERSTONE OF MARITIME HOMELAND DEFENSE

Naval War College Review, Autumn, 2003

Guy Thomas

Among the many lessons "9/11" has taught is that the United States is a vulnerable nation. This is especially true on its sea frontiers. President Franklin D. Roosevelt understood this; he made a point of it during his first "fireside chat" after Germany invaded Poland, plunging Europe into war in September 1939, twenty-seven months before the U.S. Navy was attacked at Pearl Harbor. American security was, he said, "bound up with the security of the Western Hemisphere and the seas adjacent thereto." It still is. "We seek to keep war from our firesides by keeping war from coming to the Americas." Today, we are engaged in a different war, one that has already come "to our firesides." To help prevent its return Americans must again attend to the security of the seas and their ports. This is doubly true for, despite the emergence of the information age and the decline of the U.S. merchant marine, the United States is still a maritime nation; the security of its harbors and seaports is still of first importance to the well-being of this country. Americans are very dependent on maritime trade, as was recently demonstrated by the significant economic damage done by the short dock strike on the West Coast. It is easy to envision that the economic cost and social impact of simultaneous terrorist attacks on two or more American ports would be huge.

The nation is attempting to grapple with this problem, which is ultimately one of global scope. One part of that problem—but a step that is both critical and manageable in the short term—is to maintain the security of its ports. The United States needs to track and identify every ship, along with its cargo, crew, and passengers, well before any of those vessels and what they carry enter any of the country's ports or pass near anything of value to the United States. This article proposes a system that would provide that tracking capability and meet any related emergency with an appropriate response. This proposal—the result of months of war games, conferences, and working groups dealing with the maritime aspects of homeland security—is intended to be a strawman, a thought starter, a means of generating informed debate on how and why the United States might build a maritime counterpart to the flight-following systems of the North American Aerospace Defense Command (NORAD) and the Federal Aviation Administration (FAA).

Not everyone supports this idea. Some believe it is too difficult, or not worthwhile, or both. Admiral Vern Clark, the Chief of Naval Operations, is not one of these; he has twice called for the creation of a "maritime NORAD." He first urged its creation on 26 March 2002 during a conference on homeland security issues sponsored by the Coast Guard and the Institute of Foreign Policy Analysis at Cambridge, Massachusetts.

Parts of his speech resemble an early version of the white paper this article is drawn from, written by the author, and forwarded to the Navy Staff in November 2001. Other powerful members of the U.S. government also spoke, but it was Admiral Clark's words that the press highlighted. The CNO's second call for a maritime tracking system came on 15 August 2002, at the Naval- Industry R&D Partnership Conference in Washington, D.C. This time the press missed it:

In conducting homeland defense, forward deployed naval forces will network with other assets of the Navy and the Coast Guard, as well as the intelligence agencies to identify, track, and intercept threats long before they threaten this nation.

I said it before, and I will say it again today: I'm convinced we need a NORAD for maritime forces. The effect of these operations will extend the security of the United States far seaward, taking advantage of the time and space purchased by forward deployed assets to protect the United States from impending threats.

What, some ask, does the Admiral mean by "forward deployed assets?" If he means units deployed overseas, the problem is significantly more difficult than if he means units under way (in fleet operating areas, for example) a few hundred miles off the U.S. coasts. A maritime NORAD-like system could be built from existing technology to solve the "detect, ID, track, and interdict as appropriate in the coastal-belt" problem. That belt could extend from fifty to a thousand miles offshore, or some other similar area, to provide sufficient time for early detection, analysis, and determination of the threat potential or the probability of involvement in illegal activity of vessels enroute to the United States. A maritime traffic tracking system as outlined below would require almost no additional tactical assets and would make the ones that are there substantially more effective. The overseas, far-forward problem is a closely related, but separate, issue. Its multinational political dimension alone makes it substantially more difficult. However, it is not significantly more difficult technically, once we get foreign ships in foreign waters to install the proposed transponders, which, it must be Admitted, would indeed be a very tough sell. The proposed system would, most assuredly, assist in the forward-deployed situation, but, in any case, the problem of security at home needs to be solved first. It may be possible to expand overseas the tracking capabilities required once they are in place in U.S. coastal waters and economic exclusion zones, but it would be nearly impossible to do the reverse—to establish the required tracking capabilities in foreign seas and then extend them back to the coast of the United States. To attack the overseas environment before the near-home coastal problem would result in a huge waste of time and national resources, both manpower and money, and would leave our ports still vulnerable.

THE PROBLEM

The United States has 185 deepwater ports. Every day over two hundred commercial vessels and twenty-one thousand containers arrive at eighteen of these deepwater ports. The container-carrying ships are largely concentrated in less than a dozen ports that have the proper handling equipment, but most ports can accept a few containers. Additionally, approximately five thousand vessels of all types, pleasure boats, fishermen, tugs with or without tows, oilfield-support vessels, and research ships are active every day in the vast area from fifty to a thousand nautical miles offshore. All these vessels are large enough to carry significant cargoes. They sail to and from not only the 185 ports mentioned but also an even larger number of smaller moorings and anchorages. Some of these vessels, which are of all sizes and types, are involved in illegal activities, such as drug and immigrant smuggling, illegal fishing, or environmental pollution.

The concern since "9/11" is that there may be other vessels with even more sinister objectives. This concern is heightened by the fact that tens of ocean-crossing-capable commercial vessels disappear every year. Some sink because of weather or unseaworthiness. Others probably "disappear" for insurance purposes. More than a few are attacked by pirates. Additionally, older but serviceable ships of considerable size can be purchased in many places for less than the terrorists probably spent to execute the attacks on the World Trade Center and the Pentagon. Any of these vessels could carry enough explosives to destroy or substantially damage a port's infrastructure, including bridges, chemical and petroleum plants, processing, handling and storage facilities, and such high-value vessels (and thus high-payoff targets) as aircraft carriers and liquid natural gas carriers. Indeed, the easiest way to put a weapon of mass destruction into large urban areas such as New York, Los Angeles, or the Hampton Roads area of Virginia is to send it by ship. A relatively small explosion onboard a small ship with a deck cargo of even a few smallish bags of anthrax or some other evil substance in a major city port might only kill a few thousand or even just a few hundred people, but the terror it would cause would be devastating to our economy, if not our national psyche. The threat to our ports is especially true now that the airport and container security has been significantly enhanced worldwide.

These facts make it apparent that the United States needs a better means than it now has of identifying and tracking all vessels, as well as their cargoes, crew, and passengers, as they approach the coasts of the United States or its territories. The country does not now have a system that will give full "situational awareness" of the surface of the seas surrounding it. It needs to create one now. We need to know the name and ownership, position, course, speed, and intended port of call of every vessel; the identity of everyone onboard; and a description of its cargo or function—just as is required for all aircraft, private and commercial alike. In other words, what is needed is a requirement for a "float plan" (the maritime equivalent of an aviation flight plan) and a means of positively identifying each vessel well before it nears our coasts (e.g., the maritime equivalent of an "identification friend or foe," or IFF, system). Moreover, the float plan and the maritime IFF system must be linked together. Such an infrastructure might be a "North American Maritime Defense Command." Various proposals are under investigation by the governments of the United States and Canada via a Bi-National Maritime Awareness and Warning Working Group based at NORAD. Others have suggested changing NORAD to the "North American Defense Command" (with the same acronym), with air, land, and sea components.

HOW CAN THIS BE DONE?

Once we have a workable long-range maritime IFF, we can

use several existing technologies to gather, process, analyze, and fuse data from all useful sources so those who must daily make decisions can reliably make the right ones in a timely manner and take appropriate action. As already noted, the proposal centers on a maritime analog of the FAA and NORAD, as well as the

U.S. Customs flight-following systems and the development of a long-range maritime IFF. This is the critical initial step in building a maritime equivalent of NORAD. Though it does not address adequately the very difficult problems of tracking the cargo and the people onboard, this increment will provide an "information backbone" with which data on the contents of a vessel—its cargo, crew, and passengers—can be melded, as it absolutely must be. Though this article focuses primarily on a maritime IFF system and the needed information backbone, it also addresses the other issues, i.e., the gathering, processing, analyzing, fusion, and provision of data, to provide a context and to outline issues to be considered for an end-to-end "system of systems."

The tracking of ships bound for the United States is a task for the U.S. Navy, U.S. Coast Guard, and U.S. Customs Service. Whereas ship tracking is now undertaken only by exception, when extraordinary circumstances warrant, this article proposes that it be done on a routine basis. Indeed, given today's technology, its comparative low cost, and substantial capabilities, it would not be excessively expensive to put a transceiver or transponder on every ship and track it (as will be discussed below). However, even if a transponder could be placed, at a reasonable unit price, in every container bound for the United States, the aggregate cost could well prove prohibitive. But the payoffs of even just vessel tracking for the struggle against terrorist threats (as well as drug and illegal- immigrant smugglers and polluters) could be substantial, far outweighing its cost.

Surveillance under the proposed system would be focused on the belt from fifty to a thousand miles offshore, or some other similar zone. (Vessels on voyages originating and terminating within U.S. waters would be of interest only if they ventured more than fifty miles offshore.) Vessels in that belt would be forbidden to approach U.S. shores closer than twelve nautical miles (the international recognized limit of territorial waters) without having switched on and operated a maritime IFF system (any of the several described in the appendix) for at least the previous ninety-six hours. A ship departing a foreign port less than ninety-six hours from the coastal waters of the United States would have to have the system operating as it gets under way.

Also, all vessels bound for U.S. waters would be required to file a float plan (with the information detailed in the sidebar) and have a registration receipt from the U.S. Coast Guard before reaching a point ninety-six hours (about a thousand miles, at ten knots) out. Those who did not comply would risk being stopped, searched, and denied entry to U.S. ports for a minimum of two days. The float plan could be forwarded via e-mail or any other record-producing communications system. Most shipping companies already do something like this internally to keep track of assets and maintain business flow. This is an expansion of the field of vessels for the Advanced Notification of Arrival (ANOA) now required by the Coast Guard for large vessels entering our ports. It is in any case a good idea from a safety view, as a float plan tells someone ashore where a vessel is headed and when it expects to get there; if the vessel does not arrive on schedule, a search can be initiated. (In two recent cases, men sailing alone spent more than three months adrift in disabled boats because no one knew to look for them.) Many smaller ships operating offshore already have communications devices that support e-mail; those that do not could use a marina's email before departure. It would, in any case, be the operators' responsibility to make the necessary reports and to obtain the necessary documents. Given the widespread availability of communications systems, however, this requirement should not be arduous. The cost of the transponders and the minimum monthly fee for U.S.

citizens could be funded with an income tax credit. In that most of the proposed transponders would also have at least email capability, additional usage of the system would be the vessel operators' responsibility, like exceeding a monthly allocation of cell-phone minutes.

These reporting requirements are consistent with international practice regarding freedom of navigation on the high seas. Indeed, the U.S. Coast Guard already has a ninety-six-hour Advanced Notification of Arrival requirement in effect, dictating that large commercial vessels broadcast their intentions well before they cross the thousand-mile line. Once within a thousand miles the proposed maritime IFF system would update a vessel's position at specified intervals as it closed the coast. The fifty-nautical-mile inner boundary eliminates from surveillance the vast majority of pleasure and fishing boats and other coastal commercial vessels that normally do not routinely venture far offshore. The boundaries, both far and near, could be easily adjusted as needs and experience dictate.

Those areas that abut neighboring countries' borders will need special attention, including the establishment of radar identification zones. The areas include where the coasts of Texas and California meet Mexico; where Washington State and Maine meet the Canadian coast; the Strait of Florida, which abuts the territorial waters of Cuba; and the vicinity of Puerto Rico and the U.S. Virgin Islands. Radar surveillance in these high-interest, potential high-threat areas would greatly facilitate the positive identification of all maritime traffic, especially if very-long-range (110 nautical miles–plus) high-frequency surface-wave (HFSW) radar is employed. Indeed, means are already at hand in most of those places to provide the close surveillance required. The one thing they are lacking is the means to identify positively the many tracks they now have. This proposal solves that problem for the tracking of all law-abiding citizens. The others would become much more conspicuous. Where adequate radar surveillance is not now available, a few well-placed aerostats, like those used in counterdrug operations, would provide sufficient coverage. However, experience indicates that radar tracking is not enough—satellite communications transponders onboard ships, serving an IFF function, are key to solving the ship-traffic management system.

SHIP AND CONTAINER TRACKING

Monitoring the contents and tracking the location of containers are at the heart of shipping security. Many people believe containers, whether arriving by land or sea, represent the greatest potential for security breaches and entry of contraband. The tracking of containers bound for the United States is an important responsibility of the U.S. Customs Service. The U.S. Border Patrol, Drug Enforcement Agency, and Federal Bureau of Investigation, plus other law enforcement agencies, support Customs in this effort. The people-vetting and tracking problem is even more difficult, and these agencies also assist the Immigration and Naturalization Service (INS) in vetting and tracking the people arriving in the United States via all modes of transportation, including ships. (For some of the currently available technologies, see the appendix.)

Potential solutions to these two problems will not be addressed here other than to note that the float plan, systems, databases, and procedures developed to track ships would assist the INS in its people-tracking efforts and the U.S. Customs Service in its cargo-tracking mission as well. In fact, the system proposed here would have much wider applications than port security, or even counterterrorism generally. As a start, it would also greatly assist in the war on drugs, help curb illegal immigration, assist in fisheries protection, and support antipollution operations.

DRAFT NOTICE TO MARINERS

Be advised: All vessels intending to enter or transit the territorial waters of the United States or its protectorates (Guam,

Puerto Rico, Virgin Islands, Samoa) must file the Advanced Notification of Arrival (ANOA), as required by pertinent U.S. Coast Guard regulations, or a float plan as described below with the U.S. Coast Guard, prior to arriving within one thousand nautical miles of the coast of the United States or its protectorates. If the point of departure is within [to be specified] nautical miles, the float plan must be filed a minimum of twenty-four hours prior to leaving the foreign port. The float plan will include:

1. The names and nationalities of all persons onboard
2. List of all Maritime Mobile Service Identifiers (MMSIs) to be used on the voyage.
3. Description of all cargo
4. Point of last departure
5. Destination
6. Estimated time of arrival
7. Estimated time and location of arrival at a point fifty nautical miles from the coast of the United States or its protectorates.

Additionally, all vessels must also have one of the following systems on and transmitting its identification (MMSI) and location. It must be reporting the vessel's position and MMSI not less than once an hour when in international waters within [to be specified] nautical miles from the United States or its protectorates. When in international waters within [to be specified] hundred miles of the United States or its protectorates and planning on entering U.S. territorial waters the vessel must broadcast its identification and position four times an hour. Vessels not complying with this directive will be subject to interception and detention for a minimum of twenty-four hours at the limits of U.S. territorial waters.

Most of the civilian agencies named above already have at least limited maritime surveillance capabilities to cope with such problems. As an example, the Customs Service has an excellent facility at March Air Reserve Base, near Riverside, California—the Air and Marine Interdiction Coordination Center (AMICC). It is primarily focused on countering air smugglers and tracks all aircraft crossing any border in North America. The coverage of the remote radars (displayed at AMICC via live video feeds) extends far across North America and well into South America. AMICC currently makes only a minimal effort against marine smugglers, due to manpower and equipment limitations, but Customs would like to see that capability expanded. The agency clearly understands what needs to be done and given the resources, is ready to do it or to help whatever other organization gets the job.

At any one moment there are about five thousand aircraft airborne either over the United States or in its immediate vicinity. The Customs Service's system for coordinating multiple reporting entities and the tools it has developed for its air surveillance task are especially instructive. In the course of a day, seven to eleven AMICC watch standers routinely select an average of 2,900 tracks (out of tens of thousands) for special, detailed examination. To assist in that examination AMICC has developed an excellent set of software tools that allow surveillance system operators to access databases that contain the current flight plan data and the flight tracks of all flights of the aircraft under special scrutiny in the past two years, as well as data on anyone of special interest who has been associated with that particular aircraft. Interestingly enough, the Coast Guard has much of this same data on over six hundred thousand vessels of the U.S. registry in its Maritime Information for Safety and Law Enforcement (MISLE) database. It also has many of the same types of interfaces to a host of other organizations, such as commercial insurance databases and international police organizations as does the AMICC. The AMICC is also a major participant in the Domestic Events Network (linking the Federal Aviation Administration, NORAD, law enforcement agencies, and air traffic control facilities). AMIC's experience should prove very valuable in developing a maritime counterpart. If the maritime surveillance organization is not collocated at the AMICC, it would need to have a close interface with AMICC and be a major participant on the Domestic Events Network. The maritime tracking center would need to be linked to the MISLE database, which would need in turn to be interfaced to the Global Command and Control System, which is now under consideration, in order to approximate what is now in operation at the AMICC.

MODELS AND FRAMEWORKS

Fundamentally, the maritime homeland security/defense mission involves a *detect-assess-act* cycle. These cycles can be approached in several ways. The most famous model is the "OODA loop," which consists of the elements *observe, orient, decide,* and *act.* Another widely employed model is the "sensor to shooter" paradigm. A third, more recent breakdown of this cycle is the "find, fix, track, target, engage, assess" model. Though each of these models is useful, none fully describes what actually happens in a systems sense. Let us use a slightly different model to describe a vessel- tracking system and its interfaces with a decision-making apparatus so as to produce a system able to take timely action against potentially hostile vessels and to apprehend others engaged in illegal activities.

This model, called "Warfare in the Fourth

Dimension," was developed more than twenty years ago to describe and analyze the importance of time for decisions in combat.[3] It was first used to equate the battle for control of the electromagnetic spectrum with the battle for time, the fourth dimension in physics. The model's components are the sensors (S), the processors (P), the fusion system (F), the decision maker (DM), and the action taker (AT), as well as the communications links that tie each of those components together. The paradigm closely mirrors what actually happens in all forms of combat, be it an infantry man fighting in very close combat or a ballistic-missile-defense action on the edge of space.

Sensors detect phenomena given off by potential targets and forward data to processors, which feed information to the fusion system. The fusion system provides knowledge to the decision maker. He, in turn, takes all other factors of the environment, including rules of engagement, force status, strategic situation, political alignments, and so on, into account and develops as clear a tactical picture as possible and (ideally) the wisdom applied to it. On this basis the decision maker issues orders to the action taker. The sensors detect the results, or lack thereof, and the cycle starts all over again. A shorthand of the model's operation is S-P-F-D-A.

In close ground combat, eyes, and ears (and hands and noses, if the conflict is very close indeed) are the primary sensors. The processors, fusion system, decision maker, and action taker are all represented within soldiers, and the communications systems are the synapses in their brains. At the other extreme, on the edge of space, the sensors might be infrared or electronic intelligence satellites, linked to their processing centers on the ground by high-capacity data links that are in turn linked to the fusion system via military satellite communications or fiber-optic cable. The fusion systems might, or might not, be collocated with the decision maker. Most likely the decision maker would be linked to action takers via a separate military satellite communication system. Battle damage assessment uses exactly the same systems, tasked to look for confirming phenomena, after which the S-P-F-D-A process starts all over again.

The requirements for an enhanced tracking system are being widely discussed within the Navy, Coast Guard, and Customs. The basic requirement for overall situation awareness is "maritime domain awareness," analogous to the airspace awareness afforded by the FAA's, NORAD's, and Customs' flight-following systems. Numerous war games and conferences indicate that various existing systems could be modified to provide the basic building blocks for a system to provide the necessary awareness; this would be the first step in building a North American Maritime Defense Command. Stepping through each of the segments of the S-P-F-D-A model, let us examine how this could be done.

Sense

The first step in this chain is to select specific phenomena that can be detected by sensors and processed by the rest of the cycle in a timely manner. This is the heart of the proposal. Beyond the traditional sensors, such as radars, signals intelligence, and acoustic devices, there already exists a set of cooperative reporting systems, communications satellite–based identity and position reporting systems—the InMarSat, ARGOS, and OrbComm, communications satellite systems with midocean coverage—each of which could be adapted for use as a primary sensor for maritime domain awareness. GlobalStar and Iridium communications satellite systems, the only two other systems with similar coverage, are also developing similar transceiving or transponding systems. Yet other companies, Comtech Mobile DataComm and Boatrac as examples, have developed transceiver-based unit- tracking systems that could possibly participate in the envisioned system. Other satellite communications– associated companies and systems probably would also be able to provide basic components of a maritime IFF system.

These systems would need a common vessel-identification scheme, and one is readily available. Several of them already use the Maritime Mobile Service Identifier (MMSI), assigned by the International Telecommunications Union. Discussions with developers of most of the other systems indicate that their systems could be relatively easily modified to broadcast an MMSI as well.

If the envisioned MTTS transponder system is the maritime equivalent of the aircraft system's IFF, the MMSI is the specific entity's identification ("squawk") code. It would become an "electronic license plate." Aviation IFF was originally interrogated solely by military radar systems, but now it is the primary electronic means of identification of radar tracks for both civilian and military uses. Radar is the vital part of the IFF system, interrogating unit-based transponders and reading responses. However, a ship-tracking system such as would be required for a maritime defense command would need to track ships well out beyond land-based radar ranges; communications satellite transceivers and transponders would serve in its place. Of the five communications satellite systems that either now or would soon be able to meet the reporting requirements over a broad ocean area, InMarSat and OrbComm appear able to provide timely position reporting with oceanic coverage. As of early 2003, two other satellite communications systems, GlobalStar and Iridium, were on the threshold of the needed capability. The fifth system, ARGOS, has an oceanic communicating and reporting capability but has significant built-in time-lateness. Additionally, once a firm market and a known requirement exist, other satellite companies may well decide to provide the required services, either by adapting existing satellite systems or by including oceanic capability in new ones. (Brief descriptions of the MMSI and the several satellite tracking systems suitable for maritime use are in the appendix.)

Process

The signals containing the unit's identification and location would be broadcast via a transceiver or transponder onboard every ship desiring to enter the coastal waters of the United States. The signal would be received by one of several communications satellite systems, depending on which transceiver/transponder was installed. Overall course and speed would be calculated at the terrestrial tracking station.

Eventually, the effort could include the Automatic Identification System (AIS)—an excellent, high-fidelity collision-avoidance and traffic management system now coming into use (see the appendix)—if its transponders were placed in orbit, as has been suggested, or a method were found to route the AIS signal through one of the existing communication satellite systems. The advantages to global shipping control would be significant. However, no satellites now in orbit can receive or process the AIS signal, and it is unclear when, or even if, AIS transponders themselves will be put in space. Manned or unmanned aircraft and aerostats could also be equipped to monitor AIS and used in a surveillance/patrol role, but a space-based approach might well be significantly less expensive.

Earth stations receiving the downlink transmitted by whatever satellite system would forward the generated ship-position data to both the National Maritime Intelligence Center and Coast Guard regional reporting centers of some type.

The functions of regional reporting centers could be served by the two Maritime Intelligence Fusion Centers (MIFCs), one on each coast, recently created by the Coast Guard with assistance from the Office of Naval Intelligence. Also, the Defense Information Systems Agency is experimenting with a concept it calls "Area Security Operations Command and Control" (ASOC), by which a communications and software suite would link many of the organizations involved in homeland security. The MIFC will be linked to Joint Harbor Operations Centers (JHOCs), which will use the ASOC to link to military and other government agencies—for instance, the Coast Guard, the Customs Service, the Drug Enforcement Agency, the Border Patrol—in its area of responsibility. It would be responsible for tracking all vessels in its area, assisting in assessing all contacts and deciding whether a response is required, and orchestrating any tactical response required. It would be assisted by NMIC's civilian merchant ship section, which is the organization responsible for performing long-term trend analysis as well as maintaining a daily maritime intelligence watch worldwide.

Fuse

All-source intelligence fusion would primarily take place at the NMIC, but the MIFCs and battle watch organizations maintained at numbered fleet headquarters would assist. Coordination would be over SIPRNET (the U.S. government secure Internet), but because much of the data is not classified, the World Wide Web could also be used. Data from national collection means, including signals intelligence and acoustic systems, over-the-horizon radars such as the ROTHR* and HFSW systems, sighting reports by Navy, Coast Guard, and Customs vessels and aircraft, human intelligence, and acoustic sensors would be melded with the transponder-supplied positions to determine the presence of non-reporting vessels or tracks displaying abnormal behavior or with suspicious histories.

This is not an insurmountable task. As mentioned above, the Customs Service's Air and Marine Interdiction Coordination Center investigates an average of 2,900 anomalous tracks daily. Careful analysis and prompt information exchange with other governmental agencies and with private entities clears the vast majority of unusual tracks, but almost every day the AMICC initiates intercepts by Customs aircraft. Similarly, in a maritime defense system, Coast Guard or Navy assets, either air or surface, could be dispatched to interdict, interrogate, and determine the status and intentions of the few entities judged sufficiently suspicious by the regional reporting center—vessels not reporting or reporting in anomalous ways (such as using the MMSI of a ship known to have been recently in another part of the world). The patrol units would be linked via UHF satellite communications to the MIFC, which in turn could access the vessel's "master file" (probably at the National Maritime Intelligence Center). The master file would contain everything known about the vessel and its owners, including type, the current float plan, and all previous ones, associated MMSIs, history of ownership, and car goes and crews, plus any special notes that have been appended in the past, such as association with suspicious entities or activities.

The patrol unit, which could, in many cases, also determine a vessel's Maritime Mobile Service Identifier via a standard marine VHF radio equipped for Digital Selective Calling, would query the vessel database using the MMSI, much as a highway patrol officer runs a license plate check. A query to a Department of Motor Vehicles database can tell a patrol officer if a suspicious car should be pulled over; an MMSI check would provide the same benefit to maritime forces. Establishing the MMSI as an IFF-equivalent, electronic license plate, would be of substantial benefit to Navy, Coast Guard, and Customs patrol units. Of course, complications arise when a Navy unit must check out a suspicious entity and "pull it over"; the Posse Comitatus Act of 1877 constrains the Navy's actions in such a situation. That whole issue is under review, however, and in any case legal means can be found to halt a suspicious ship on the high seas. The USA Patriot Act of 2002 at least allows military platforms to collect intelligence on civilian entities in the manner described here.

Establishing the MMSI as electronic license plates and developing the means to track them would be important steps and would fill a substantial void in the nation's maritime defenses. Getting all units approaching the coast of the United States and its territories to broadcast their MMSIs and position is a different matter, one that would require cooperation. However, the U.S. government can require all vessels desiring to enter U.S. ports to commence broadcasting their MMSIs, within either a specified distance of the coast or time of entering port. Vessels complying would enjoy the greater safety that accrues from track following. Any ship not filing a float plan or broadcasting its identifier and location (which should be immediately obvious to patrolling units) would be subject to interception, inspection, and the likelihood of significant delays in entering port, if indeed they were allowed to enter port at all. Thus, the incentive to comply would be substantial. Delay costs all vessel owners, especially shippers, money—more money than acquiring communicating systems (that their ships should already have anyway, for safety, as discussed below) would cost them.

The processing system outlined above is an expansion of capabilities already in place at the Joint Inter Agency Task Force facilities on both the east and west coasts of the United States and at the AMICC. Fortunately, software tools in use at the AMICC and at other government agencies such as the National Security Agency and the National Reconnaissance Office have shown that the manning requirements for a full maritime watch can be quite small. New-generation display and decision technology—such as the Anti-Air Defense Commander (AADC) system developed at Johns Hopkins University's Applied Physics Laboratory, with easily understood symbology and embedded reasoning and data manipulation capabilities, now being deployed on Navy command ships and cruisers—could be used to help the regional reporting center gain and maintain situational awareness. The reporting center's display and decision system would be the focus of the data fusion efforts, such as "smart agents" (discussed in detail in the appendix), software that would sort the huge amount of data flowing in. The envisioned system could also manage communications links into and out of the several reporting and analysis centers.

Decide

A correct decision requires a sufficient quality and quantity of information and enough time to fuse that information so as to develop knowledge and then wisdom. Timeliness dictates that decision makers be able to know when they have the information—from all sources and addressing all aspects of the problem at hand, such as status of own forces, rules of engagement, and the political, strategic, operational, and tactical situations—needed to develop wisdom and issue the appropriate orders. This is by no means a trivial task; indeed, integrating vast amounts of data from heterogeneous sources is daunting for the human mind; fortunately, however, several software tools are now available to help the decision maker.

One of these is the Architecture for Distributed Information Access (ADINA) tool developed at the Johns Hopkins University Applied Physics Laboratory— an agent-based architecture for seamless access to and aggregation of heterogeneous information sources. Maritime defense regional reporting centers would use smart-agent tools like ADINA (and Control of Agent- Based Systems, or CoABS, grids, described in the appendix) both to fuse the data, including the crucial MMSI reports, and to formulate decisions and courses of action, all in close coordination with the U.S. military command structure in the appropriate area.

Act

Once the decision is made to interdict a specific vessel, an on-scene commander would be designated; rules of engagement need to be in place and clearly spell out which federal agencies would take the lead in anticipated cases. Forces, possibly including surface and air elements of the Coast Guard, Navy, or Air Force, would be assigned to take appropriate action. Rapid response would be crucial in some situations; for that reason, interdiction forces should include such regular and reserve assets as Air Force A-10s and Navy P-3s, equipped and trained for antishipping attack. Their weapons should include optically guided missiles such as Penguin and Hellfire, to allow disabling fire to be focused on the bridges and rudders of rogue ships attempting to enter port with clearly hostile intentions. In extremis, such as the need to stop a ship known or strongly suspected of carrying weapons of mass destruction, larger weapons, such as Maverick or Harpoon, must be readily available to sink it. If more time is available and forces are in position, surface units could affect the interdiction. Helicopter insertion of special operations forces, or specially trained units is also a possibility.

Navy, Coast Guard, and Customs vessels and aircraft routinely operating off U.S. shores would not only report all surface vessels in their areas but act as "first responders." Their reports would be fed into vessel master files and automatically matched with the pertinent float plan. Non-Reporting of suspicious vessels would be marked for follow-up.

Because other systems, such as InMarSat-C, AIS, and DSC (described in the appendix), broadcast position and identification information, it would be beneficial if maritime patrol forces could monitor them. Any vessel in a patrol unit's vicinity broadcasting on these internationally mandated systems could be quickly and accurately identified, by MMSI. Indeed, all units of the U.S. government assigned to surveillance and interdiction roles should also be equipped to monitor them, if not fully participate.

WHAT WOULD BE REQUIRED?

Putting this proposal into practice would require prenotification of the International Maritime Organization (IMO) but not necessarily its approval. The initial implementation of this system would require the wide promulgation of a notice to mariners directing all vessels out to a thousand nautical miles off a U.S. coast and desiring to enter American territorial waters to broadcast their identification and location at set intervals over one of the approved systems. It would further direct every vessel to broadcast its location as soon as within ninety-six hours of arrival in an American port or whichever happens first. A vessel departing a port less than ninety-six hours out would operate the system as soon as it is under way.

One final word on available technology. The International Maritime Organization already requires units above three hundred gross tons to carry InMarSat- C, as part of the Global Maritime Distress and Safety System and in accordance with the Safety of

Life at Sea Convention. InMarSat-C has a built-in ship-polling capability that meets the requirements for a maritime IFF system. The proposed system would provide that capability, all the way down to the smallest vessel capable of open-ocean navigation. These vessels will also be required to have the more expensive and more sophisticated Automatic Identification System by 2004. The purposes of this proposal could be met by either system; in any case, AIS, once it is capable of being monitored from beyond line of sight, may well become the specified system. However, AIS is significantly more expensive than the transponders of the low-earth orbiting satellite communications systems. Those other satellite communications reporting systems that would be suitable include OrbComm, GlobalStar, and Iridium. In any case, installation could be encouraged via a tax credit for American vessel owners. For foreign owners, the cost of entering U.S. waters will indeed increase, but not by an unbearable amount. Operational tests would be needed on each of these systems to ensure they are sufficiently timely and compatible with a national reporting standard. The task is clearly feasible from a technology viewpoint.

A two-tiered tracking system could be quickly emplaced, in which a combined automatic identification and satellite transceiver system sends tracking output via the AMICC or other tracking center to national and regional intelligence centers for further analysis and threat/law violation/encroachment determination, in much the same way as a Federal Aviation Agency regional center tracks aircraft. Regulations would be needed requiring all oceangoing vessels to install satellite communications reporting systems and operate them within a certain distance from the United States if the vessels intend to enter its territorial waters.

It would be very much to the benefit of U.S. security, maritime and otherwise, if the system and legal requirements outlined above were enacted immediately. This proposal is intended as a point of departure for building the maritime portion of the homeland security mission capabilities package. It names specific systems, but if more capable systems become available or a more beneficial alignment of existing systems can be made, so much the better. One way or the other, let us get on with it. We are at war, and this is a known vulnerability.

NOTES

1. Drafts of this article have been circulated since November 2001 to stimulate focused, informed debate and information exchange. That information exchange has resulted in several major revisions of this article. However, more informed discussion, war games, both technically focused and policy focused, and operational experiments are needed until the concept and procedures outlined here are fully implemented. One disclaimer is appropriate: though this article identifies specific systems to provide points of departure for further investigation, it is not intended to champion any specific system or systems. If there are better, more useful systems available either now or in the near future, then those should be used.

2. For a detailed study of the piracy problem see John S. Burnett, *Dangerous Waters: Modern Piracy and Terror on the High Seas* (New York: E. P. Dutton, 2002).

3. The model was developed by the author, as a research fellow at the Naval War College, in Newport, Rhode Island.

APPENDIX:

CAPABILITIES AND SYSTEMS FOR MARITIME DOMAIN AWARENESS

Several existing capabilities and systems, mentioned in the main article, could assist in providing an offshore self-reporting tracking capability, the situational-awareness backbone for the maritime domain. To avoid detailed explanation within the article, these systems and capabilities are described below.

Maritime Mobile Service Identifier (MMSI).

The MMSI is a nine-digit "string" designed to be "transmitted over a radio path in order to uniquely identify ship stations, ship earth stations, coast stations, coast earth stations, and group calls. These identities are formed in such a way that the identity or part thereof can be used by telephone and telex subscribers connected to the general telecommunications network principally to call ships automatically" (from Appendix 43 of the International Telecommunications Union Radio Regulations, as quoted on the U.S. Coast Guard Navigation Center website). In the United States, the National Telecommunications and Information Administration assigns federal MMSIs. The Federal Communications Commission assigns nonfederal MMSIs, normally as part of the ship-station license application; an MMSI is assigned whenever a vessel purchases an InMarSat terminal, an Automatic Identification System (AIS), or a Digital Selective Calling (DSC) maritime radio. Additional information can be obtained at the U.S. Coast Guard Navigation Center website, www.navcen.uscg.gov/marcomms. Inquiries can be directed to nisws@navcen.uscg.mil.

Automatic Identification System.

AIS is an International Maritime Organization (IMO)-approved system developed at least in part by the U.S. Coast

Guard as a collision-avoidance, port traffic-control system. Its primary component is a small broadcast transceiver, broadcasting at 160 MHz in Time Division Multiple Access (TDMA) format. A VHF signal, AIS propagates only in line of sight, so it can only be used when within visual distance of another transceiver, either mobile or ashore. The IMO has agreed to require that one of these transceivers be active on every commercial ship greater than three hundred gross tons by 2004. The system provides identification (MMSI) and location, as well as range and bearing from all other units in sight, in its field of regard. This capability allows a crew to quickly calculate its vessel's closest point of approach to all other AIS units in range. AIS also provides vessel tracking and control to Coast Guard captains of the port and their counterparts worldwide. The system's collision avoidance benefits alone are such that most merchant ships always leave it on, even on the high seas, for warning should risk of collision develop. There is an international effort to establish an interface to satellite communications systems to allow long-range tracking of all large commercial vessels.

Digital Selective Calling.

DSC capability is built into new maritime VHF radios, preset as channel 70. It is configured as a search and rescue system with additional selective calling capabilities, but it does not use TDMA or any other modulation scheme that would permit it to share its frequency; thus, it quickly saturates with co-channel interference when more than a couple of units are active at the same time in the same area. This limitation precludes its use as a wide-area surveillance system. In localities, however, it is an excellent identification and tracking tool, as it employs the MMSI and can be coupled to the Global Positioning System (GPS) or LORAN, automatically providing the unit's location. (However, since the location data can also be manually inserted, this system could be used to send deliberately false and misleading position reports.)

Commercial Satellite Communications Systems.

The third set of systems of interest to a "maritime FAA" comprises the various commercial communication systems that routinely operate over water. Those systems include the several manifestations of InMarSat, a geosynchronous satellite communications system designed to provide worldwide maritime coverage; the Argos system, which operates in a highly elliptical orbit; and the low-earth-orbiting OrbComm, GlobalStar, and Iridium communications satellite systems. These systems either have now or could be easily modified to develop position reports from either an embedded or separate GPS system and transmit periodic short, formatted reports containing time, position, and MMSI reports, thereby reporting location and identification of specific units at a specific time. The several communications ground systems receive these messages from mobile units and from them generates identification, position reporting, and tracking data at user-determined intervals. Furthermore, these are two-way systems; a vessel can be selectively interrogated, and its transponder forced to send an immediate position report, change its reporting interval, or query attached sensors. (These functions cannot be modified on board.) The automatic position reporting interval can be preprogrammed in firmware of the satellite communicator anywhere from once a day to once per minute. For instance, a reporting interval for vessels at sea might be four times per day, then automatically increase to once per hour when the vessel crosses a specified threshold, such as five hundred miles from the coast. At the twelve-mile limit the interval might increase to four times per hour, and within the boundaries of a port to once every ten minutes or less. Using other preprogrammed features, the unit can report any time it is rebooted, when the vessel starts to move, stops, or slows below a specified speed, etc. A dual-mode device containing both satellite and ground-based cell-phone transceivers, as already exists with some systems, would provide redundancy as the vessel enters

U.S. waters. These systems would thus provide all the

functionality of the radar-based IFF system, but at much greater ranges, since they are satellite based.

The current systems range in cost from $450 to over five thousand dollars, depending on additional capabilities. Most of them interface to asset-tracking software that maintains a record of where each transponder is, and was at any point in time, as well as other pertinent data. The cost would most probably decrease if their use were mandated and became widespread and there was more than one system that met the standard. Indeed, no one satellite communications system should be specified; rather, a standard should be created, and all applicable satellite communications systems encouraged to build to the promulgated standard.

It is estimated that up to 80 percent of all commercial ships still employ radio teletype as their primary means of communications. Additionally, between 10 and 15 percent of all commercial ships still rely on either continuous-wave manual Morse or single- sideband voice. However, the satellite transponders are now so inexpensive that they can reasonably be required for all ships entering U.S. territorial waters, just as all aircraft except the very smallest must be equipped with an IFF.

Container Tracking Systems.

One other aspect of the various satellite-based transponder-tracking systems is their ability—already in widespread use—to track high-value/high-interest containers from point of origin to final destination. If the United States decided to put transponders on every U.S.- bound container worldwide (five hundred thousand units?), economies of scale would bring the per-transponder cost down from over five hundred dollars, acceptable only for high-value/interest shipments, to about $170, plus installation. (Installation costs vary depending on the type of container—reefer, dry cargo, etc.—and whether the unit is installed at the container's manufacture or retrofitted.) Significantly more capable transponders, which can be configured to tell if and when, and for how long, the container has been opened or out of a specified temperature range (important for fruits, vegetables, meats, etc.), will cost in the three-hundred-dollar range. In some systems, such as OrbComm, these transponders can be programmed, when within range, to switch automatically from the satellite system to the Global System for Mobile Communications (GSM) cellular telephone grid, in use throughout the world and one of the three primary systems in the United States. GSM is also useful for vessel tracking because it provides, at low cost, a high rate of position reporting should an agency desire to track a container or vessel in port more closely than is economically feasible with the satellites since, once the ship carrying the container is within line of sight of a cell-phone tower on the coast—as far as seven miles out— the reporting automatically switches to the cell-phone system, with its lower costs.

Shipping lines currently use these systems for many commercial reasons, among them improving demurrage billing, thereby reducing periods in which a container earns no revenue and the number of containers that must be kept in circulation. Some companies estimate that they can reduce container inventories by up to 15 percent. Other advanced features available to shippers (aside from security) are designed to reduce operational costs through automation, consolidation, and centralization.

Ultimately, there must be a convergence of commercial and security interests if this level of sophisticated vessel and container management is to be pursued on a wide scale. If adoption of such a hardware/service standard, already in use for commercial purposes, allowed shippers to develop a trust relationship with monitoring agencies and thereby reduce entry time into the United States, Canada, and Mexico, the unit cost of the hardware would be further justified. "Associated data" such as cargo manifests, transport routing, crew lists, etc., would be shared by agreement with foreign and domestic customs services and other government authorities, as well as with specific entities ashore, including family and exporters, importers, and shippers, as the specific user would desire. All users could determine who, besides the government, could access their specific files. With appropriate electronic customs seals on cargo containers, every shipment can be continuously tracked from the point of origin anywhere in the world to its final destination in the United States. Customs and other civilian and military authorities would be able to view the manifests as soon as the containers are sealed, and data posted electronically by the shipper. Although the form and substance of the electronic postings now vary from country to country, sufficient data is available to form the basis of the system proposed by this article. Future standardization, however, is desirable.

Additionally, satellite communicators inside cargo containers can be programmed to send a security report when the container is sealed at the point of origin, when it is loaded on a vessel, and when it is removed from the vessel. Should the container be opened after it is initially sealed, an alarm is transmitted automatically. Built-in sensors automatically report conditions within and around the containers. This cradle-to-grave monitoring is economically practical with existing hardware. The only question today is whether to retrofit existing containers or build the equipment into new units only.

Multilevel Security.

Multilevel security, the need to mesh the data exchange requirements from systems with differing levels of security, has plagued the military and intelligence communities for years and is now greatly complicating homeland security efforts, due to the increased need to tie military security systems to law enforcement classification systems. Backoffice associations (not exposed to the general public), the trust relationships required by civil and military authorities, and the hardware and software needed to effect multilevel security are all available and in commercial use today at every classification level (except special compartmented intelligence, which is and must be treated separately).

Good models already exist, such as Pennsylvania's Web-enabled statewide criminal Justice Network (JNET), recently demonstrated to Admiral James L. Loy (USCG, Ret.), formerly commandant of the Coast Guard and now head of Transportation Security Administration. JNET constitutes a virtual single system, based on open Internet technologies, that links information from the seemingly incompatible systems of sixteen criminal justice agencies. It enables agencies to share information but does not affect independent operating environments. As required by certain confidentiality statutes, each agency can determine the extent to which the others have access to its data. JNET is a secure "extranet," providing a secure publish-and-subscribe architecture featuring encryption and digital user/server authentication certificates.

Area Security Operations Command and Control.

The ASOCC is a Defense Information Systems Agency (DISA) "advanced technology concept demonstration" that combines information-handling tools into a command-and-control tool for homeland security. Conceptually, it could link all the communications capabilities envisioned in this article. It would be the link to both the NMIC and the military command authorities, as well as to the regional intelligence centers and civilian authorities, including law enforcement agencies—the Federal Bureau of Investigation, the Customs Service, the Drug Enforcement Agency, the Border Patrol, the Immigration and Naturalization Service, and the Coast Guard—and their intelligence entities. After "9/11" it was realized that no such capability existed; the ASOCC concept was the response.

DISA is also developing an overseas counterpart to the ASOCC, the Coalition Rear Area Security Operation Command and Control (CRASOC), to tie U.S. overseas rear-area force-protection assets into host-nation force-protection assets for mutual support. Its test bed is in Japan, at the headquarters for the Commander U.S. Forces Japan. The engineering design models for both the ASOCC and CRASOC are in use today, and further development is under way. The proposed maritime homeland-security center would also be linked to both the U.S. military command authority in its area and the NMIC via wideband communications channels using the ASOCC. The ASOCC would also provide wideband communications to civilian satellite communications systems, and thereby access to the databases of such organizations as Interpol (to which NMIC is already linked) and, potentially, commercial databases such as that maintained by Lloyds. These tie-ins would be crucial for tracking and identifying cargo and crew, but they could also have a major impact on efforts to identify and track suspicious vessels. The ASOCC would assist local fusion points for all information in a region and provide linkage to the Coast Guard maritime security squadrons, captains of the port, and port masters of a region, as well as to regional intelligence centers. Since November 2001, the ASOC has become the information backbone for a new force protection entity called the Joint Harbor Operations Center (JHOC); it is being used almost exactly as is postulated here, except its focus is on close-in harbor and port security.

Smart Agents.

Software tools known as "smart agents" are now being experimented with in several warfighting roles. One such effort is the Defense Advanced Research Program Agency's CoABS (Control of Agent-Based Systems). It arrays smart agents in grids to perform specific warfare-associated information-manipulation functions. CoABS grids could assist in processing data from the sensors, automatically routing items to databases while passing tasking refinements back to the sensors and forwarding to the fusion system information they derive from sensor data. The fusion system could have its own set of CoABS, passing tasking updates back to processors and sensors while routing correlated (linked together) or collated (intermixed) data on to the decision and display system.

Annex 2

Johns Hopkins Applied Physics Laboratory

Press Release of July 28, 2020

Former Johns Hopkins Applied Physics Laboratory (APL) employee Guy Thomas knows a thing or two about persistence.

In response to a direct order issued on 9/11 by President George W. Bush to the Chief of Naval Operations, the Navy established special presidential taskforce looking for ways to mitigate maritime terrorism, including ways to help the U.S. identify and track all vessels approaching the country. Thomas — then an APL employee in the National Security Analysis Department and liaison to the Naval War College — served as the technology co-lead on the effort. At that time, the existing Automatic Identification System (AIS) system worked for short-range detection, but long-range tracking was not considered possible.

A former U.S. Navy space systems subspecialist, Thomas postured that a space-based solution was the answer to offshore tracking and identification and went looking for an answer via space systems. When he received a brief on AIS "I recall asking whether anyone ever thought about putting an AIS receiver on a low-Earth-orbiting satellite," said Thomas. The answer from the five engineers in the room was the concept was not feasible. Thomas thought they were wrong and set out to prove it, with the help of his APL colleagues.

It would take another three years to sell the idea and receive funding for a prototype, and then a four year battle with a number of other organizations over many different items including with NSA which tried to classify the concept, but his years-long determination to form the Satellite Automatic Identification System (S-AIS) is now being recognized with a nomination for the <u>National Medal of Technology and Innovation (NMTI)</u>.

The prestigious NMTI award is the highest honor bestowed upon individuals for technological achievement in the U.S. Presented by the president, the medal is awarded to innovators for their outstanding contributions to America's economic, environmental, and social well-being. The purpose of the National Medal of Technology and Innovation is to recognize those who have made lasting contributions to America's competitiveness, standard of living, and quality of life through technological innovation.

"I'm very honored to receive this nomination for work that I started while at APL," said Thomas. "My network of APL colleagues were a critical element of support in both the development of the hypothesis, and the dogged pursuit which finally resulted in the creation of S-AIS. It probably would not have happened without them."

Thomas' work on S-AIS has helped transform maritime safety and security operations globally. No longer just a concept, S-AIS is now the primary ship identification and location system for vessels world-wide. Its information is widely used as the core of maritime domain awareness, search and rescue, environmental monitoring, resource protection and maritime intelligence application such as counter-smuggling and sanctions violations detection.

But back in 2001, S-AIS was a hard sell.

Thomas faced hurdles at every turn. He had to convince stakeholders of the value of the system and that it could actually work. There were doubts that a signal could be picked up from space because of the density in that portion of the frequency spectrum. Some simply thought the idea was not worth the cost.

Thomas eventually secured enough funding through the U.S. Coast Guard and negotiated a public-private partnership with ORBCOMM, a satellite communications company, to build an initial prototype. The first six commercial S-AIS satellites were launched in 2008. S-AIS has since helped the U.S. Coast Guard and navies and coast guards all over the world observe approaching ships and vessels and has also become a primary tool for the protection of the maritime environment and its resources.

"[Thomas'] persistence in fostering collaboration and pursuit of S-AIS dual use for both commercial and government sectors resulted in a practical solution for maritime logistics and a national security layer against non-cooperative vessel threats," former U.S. Deputy Undersecretary of Defense John Kubricky stated in a letter nominating Thomas for the award.

"Few technologies have had such a significant impact on the vitality of this nation's economy as has S-AIS, which will remain the dominant baseline for decades of maritime operations in the future."

Annex 3

Naval War College Review

Volume 36
Number 1 January-February

Article 2

1983

Warfare in the Fourth Dimension-Is the Navy Ready for it?

G. Guy Thomas

Follow this and additional works at: https://digital-commons.usnwc.edu/nwc-review

Recommended Citation

Thomas, G. Guy (1983) "Warfare in the Fourth Dimension-Is the Navy Ready for it?"
Naval War College Review: Vol.36 : No.1, Article 2.
Available at: https://digital-commons.usnwc.edu/edu/nwc-review/vol36/iss/1/2

This Article is brought to you for free and open access by the Journals at U.S. Naval War College Digital Commons. It has been accepted for inclusion in Naval War College Review by an authorized editor of U.S. Naval War College Digital Commons. For more information, please contact repository.inquiries@usnwc.edu.

Warfare in the Fourth Dimension — Is the Navy Ready for it? How can the Navy Prepare for it?

by
Lieutenant Commander G. Guy Thomas, US Navy

Most dictionaries define the fourth dimension as time, however, some military writings refer to a fourth dimension of military endeavor beyond the three traditional ones of sea, air and land. That dimension is the electromagnetic spectrum. The two concepts may, at first glance appear to have little in common with each other but, in fact, they are closely related. The battle for the electromagnetic spectrum is, in many respects, a battle for time.

RAdm, Albert A. Gallotta, a recent Director, Electronic Warfare and Cryptology, Office of the Chief of Naval Operations, made the point very clear indeed when he said, "I view all battles as having a time line that can be divided into four phases: 1) the force coordination phase, 2) the surveillance phase, 3) the targeting, weapons direction phase, and 4) the weapon's fusing and impact phase, i.e., the actual engagement. Electronic warfare is paramount in the first three phases. "He went on to say that we need to get our operational commanders to spend more time thinking about the first three phases of the battle, rather than spending all of their time thinking about only the last two (targeting and engagement). [1]

The purpose of this paper is to define a need within the Navy for additional thinking about "warfare in the fourth dimension" and to suggest how electronic warfare could be simulated for education and training purposes at the Naval War College to add to the readiness of our Navy. Fleet readiness is the primary mission of the Naval War College and its Center for War Gaming and it must be remembered readiness is a primary requirement for any strategy involving the

Navy, no matter what its objective.

The Naval War College is acquiring a new computer-driven war gaming system, the Naval War Gaming System, to assist in maintaining that readiness. The Naval War Gaming System will have the capability to simulate EW and C^3 (command, control and communications) systems; but, as delivered, that capability needs further refinement in order to accurately play EW and C^3 systems. Currently the Naval War College inputs EW into its war games by modifying the probability of kill (pk) in its battle damage assessment calculations. This method, while better than nothing, is insufficient to accurately measure the effectiveness of EW. A major portion of the effectiveness of EW is the time delays that it generates. This is true of both offensive and defensive EW systems. War gaming Battle Damage Assessment is only a terminal defense versus offensive power calculation rather than an assessment of a system's effectiveness across all four phases of the battle.

Some senior naval officers on both sides already recognize the significance of the battle for the fourth dimension. Both Admiral Gorshkov and Admiral Moorer have indicated they believe that the side which controls the electromagnetic spectrum in the next war will be the winner.[2] However, most less senior officers are not yet aware of the impact that modern electronic warfare technology can and will have on modern warfare.

The Defense Science Board made some comments in its April 1979 Summary Report that go right to the heart of the matter: "A major reason for lack of progress within the US Navy operating forces to date in achieving counter-C^3 capabilities is that relatively few naval officers understand the significance of the change in character of war at sea that current Soviet naval doctrine and electronic warfare capabilities impose. These few who are aware are generally those whose normal duties require exposure to compartmented intelligence. The Navy, as an institution, has not yet recognized this problem. *This, a principal conclusion is that basic change is required in the attitude and emphasis of U.S. naval thought regarding the role of electronic warfare in war.* Some organizational means is required to bring about that change by defining the operational concepts to ensure a whole product —not one that is strong in one aspect and weak in another. Fleet participation in the effort is essential.

"Substantial improvements in force survivability and combat effectiveness can be attained only if the various counter-C^3 measures identified in this report are thoroughly integrated with other operational initiatives.... Current arrangements for integration of operational doctrine development with the development of requisite counter-C^3 capabilities are casual and fragmented and so do not achieve the necessary integration."[3]

The Soviets appear to be ahead of us in this regard. It has been clear for some time that they plan to control the electromagnetic environment in war just as they plan to control the other three dimensions.[4] Unclassified writings on the various aspects of a doctrine the Soviets call "radioelectronic combat" (REC) are very clear. Their doctrine integrates all aspects of EW with destructive firepower to neutralize enemy command, control and communications facilities. "Soviet EW Specialists, whose skills and equipment currently exceed our own to a significant extent, seek soft kills against a comprehensive spectrum of targets: command/control and communications; radars, fire control systems; infrared and optical seekers; navigation aids; and so on."[5] Their primary mission is to introduce time delays in an opponent's execution of his war plans, be they offensive or defensive plans.[6] This is the essence of the REC doctrine.

The United States is developing a similar doctrine that they call "command, control and communications countermeasures," (C^3CM).[7] The main difference between REC and C^3CM is that the Soviet's doctrine appears to be in place and operational while ours is an embryonic state. To be sure the United States has successfully employed electronic warfare in the past but not in the integrated, coordinated manner in which Soviet writings suggest. The Soviets have obviously devoted a comparatively large share of their resources (time, manpower, money) to this concept. Given the technical capabilities of the US defense industrial base there is no doubt that we have a broad spectrum of very capable EW/C^3CM

systems and can develop more if the resources are provided. But, even as we develop these systems, the question that remains in many people's minds is—How are we going to train our operational commanders in their use?[8] Clearly we have deficiencies in the area of EW/C³CM training that need to be rectified.

Why study electronic warfare? Electronic warfare is often referred to as the "soft kill" option but in fact is a soft kill only in that it does not do the actual destroying. Successful EW can generate time delays in bringing weapons systems to bear to the point that the weapons systems are useless. This is the focal point of this paper and should be examined in more detail. Each of the three basic elements of EW: Electronic Warfare Support Measures (ESM), Electronic Counter Measures (ECM) and Electronic Counter Countermeasures (ECCM), can be directly related to that statement. Successful ESM can provide warning to allow one time to either move out of harm's way or employ one's hard kill weapons systems against approaching enemy forces. Successful ESM can provide warning to allow one time to either move out of harm's way or employ one's hard kill weapons systems against approaching enemy forces. Successful ECM can generate time delays and confusion for the enemy which will allow greater opportunities for maximizing one's own hard kill weapons engagement opportunities. ECCM can cancel or minimize time delays and confusion factors created by an opponents' ECM.

As an example, much has been written about the effect on naval warfare of the small missile that destroyed HMS *Sheffield*. However, very little has been noted in the press on how that Argentine fighter-bomber came to be in position to launch that missile. One account did mention that HMS *Sheffield* had shut down its air search radar to avoid detection from Argentine ESM-equipped aircraft. Additionally, the *Sheffield* appears not to have detected any emissions from the Super Entendard. Thus the aircraft was able to close to within missile-launch range and accomplish its mission. The failure of the *Sheffield's* ESM gear, be it because of faulty design or equipment failure, provided the *time* needed for the plane to close the ship and launch its missile undetected. If the *Sheffield* had had the opportunity, and it appears that she did not, she could have employed ECM (jamming/chaff), to degrade the Super Entendard's targeting system. Which would have delayed the launch of the Exocet or at least degraded its probability of success.

There is another possibility beyond those faulty designs or equipment failure and that is that the position of the *Sheffield* had been fixed by Argentine (or Soviet?) SIGINT/ESM and that information was provided to the fighter-bomber either before it took off and/or while it was en route to the *Sheffield*. If this is the case, then the Super Entendard, may not have needed to turn on any emitter until it was quite close to the *Sheffield,* thereby decreasing the *time* available to the British to detect and react. One way or the other, it is those kinds of risks we run if we are not up "up to speed" on warfare in the fourth dimension.

REC and C^3CM, part and parcel of that warfare, are generally conceded to have four aspects, referred to by the Defense Science Board as the four horsemen of the modern apocalypse. They are: Exploitation, Deception, Denial, and Destruction. EW plays a large role in the first three and provides targeting for the fourth. While "exploitation" and "denial" are fully covered by the traditional terms, ESM and ECM, respectively, "deception" has many more components that just its EW aspects (which are generally grouped under ECM). However, this paper will address only the EW portions of "deception."

The Soviets are known to have large numbers of devices for the exploitation, deception and denial (jamming) of emissions. Those systems are located on land, in specially configured aircraft and in surface ships (AGIs). They are also reported to have those capabilities in many, if not most, of their surface warships as well as many of their civilian-manned ships (including merchant, fishing and scientific research fleet ships). They may also have some of those capabilities in some or all of their submarines. Additionally, they are reported to have satellites with the capability to exploit emissions.

In all the Soviets pose a very significant threat to our C^3. They can exploit our C^3 for targeting and/or intelligence and they can deceive or deny large portions of it. The above is a thumbnail sketch of the problems we face. The Soviets appear to have both the capability and intent to fully exploit the fourth dimension, be it defined as time or the electromagnetic spectrum. [9]

The US is known to possess similar systems as well, but security constraints preclude an in-depth review of either side's EW capabilities in an unclassified publication such as this. Even if we have the systems we need a coordinated doctrine. One way to develop and test the needed doctrine would be through war gaming. The end result, if fully implemented and supported, would be the capability to develop and evaluate EW doctrine, strategy and tactics. The envisioned system would also provide our operational commanders with an appreciation of their EW assets and capabilities in addition to their liabilities and vulnerabilities, (the understanding of which are, according to RAdm. Gallotta, woefully deficient in today's US fleet).[10]

By preparing our operational commanders for warfare in the fourth dimension via war gaming they will be much better able to survive and win on the modern battlefield. However, at this time, EW is not adequately war gamed anywhere.[11] War gaming of EW requires more sophistication than standard simulation and analytical models which only provide "snap shot" views of iterative processes. An EW war game system would need to be designed as a logic path or a series of logic paths that conceptually mirror real world systems. With this approach, aggregated or fine-grain subroutines can be applied either at the end of logic paths and/or on the objectives of a particular game. The majority of war games at the Naval War College are, historically, "decision-making" games aimed at education and training. The concept outlined in this paper is perfectly consistent with decision-making and tactical/strategic research games such as the Global War Game and NavMat series. An example of the benefits of these kinds of games in the play of the Orange plans prior to WWII which had such a large and direct bearing on our successful prosecution of the war in the Pacific during WWII.

If we are to upgrade our capability to play EW/C^3CM in war games at the Naval War College the first two questions that must be answered are "How?" and "To what extent?" These are questions that, although simple in appearance, require much thought and analysis. As indicated above war games at the Naval War College are designed to give both students and actual naval commanders and staffs tactical/strategic decision-making experiences leading to the development and evaluation of tactical doctrine, thus the "To what extent" part of the question may be the easier question to answer. EW should be included in war gaming to the extent that it contributes to the players' understanding of modern warfare, but not to an extent that prohibits the tactical functioning of the war game scenario. It will be important to have the right mix of "soft" and "hard" kill in the game because too much "soft" kill could bring the game to a grinding halt without demonstrating anything other than the confusion that is real war even in this electronic age.

The first part of the question, "How do you play EW in war games?" has been the subject of a great deal of discussion. Both the complexity of EW/C^3CM and the high security classification of much of the specifics of EW/C^3CM are problems that must be dealt with. Numerous organizations have undertaken studies on simulation and analysis of EW and C^3 systems for either equipment design analysis or training and tactical decision-making analysis. War gaming at the Naval War College clearly falls into the second category.

Many of the procedures and computer routines that these organizations have created for their studies may very well have direct application to the introduction of a higher level of EW/C^3CM play in naval war gaming.[12] However, several of the studies were concerned with technical (as opposed to procedural) aspects of EW. As such, they are significantly more detailed than may be required to demonstrate EW/C^3CM at the level of Naval War College usages but, their lessons learned can be incorporated into war gaming at the Naval War College. Other organizations' efforts more closely approach needs of the Naval War College. As an example, while under contract to the Electromagnetic Compatibility Analysis Center, Annapolis, Md., the IIT Research Institute designed, programmed, and installed a system that integrated the dynamic aspects of EW into a computer-assisted war game to analyze the impact of EW on US Army systems, doctrine and tactics. While this system is designed to analyze the performance of tactically deployed Army communications and EW equipment as they would operate in a land engagement, the system does appear to have a core capability which may have applicability in analyzing the impact of EW/C^3CM on US Navy systems, doctrine and tactics. In brief, there are a number of organizations currently doing that may be applicable to Naval War College needs.

In order to accurately simulate EW/C^3CM play in war gaming it is necessary to fully define both opponents' EW and C^3 capabilities; Those definitions need to include lists of platforms, encompassing ships, planes, submarines, satellites and land sites, with an index of equipment that may be assigned to each platform. Those equipments should include all known C^3 and EW systems such as radars, navigation aids, and communications systems as well as all known or postulated deception, ESM, ECM and ECCM systems. Each of the systems should be described in a separate listing with its capabilities and characteristics such as sensitivity, effective radiated power and frequency range contained in tables that could be modified by a designated war game system operator. These requirements are met in many, but not all, respects by the data base already being prepared for the Naval War Gaming System. That data base will incorporate most of the operational characteristics of nearly all Soviet and US ships, aircraft and submarines.

EW/C^3CM game play requires a routine that will accommodate man-machine-man interface for a two-sided war game. It must give the players the capability to select their options, both active and passive, and then to monitor their selections(s) in terms of status and results. The machine routine should be capable of assessing player against player actions on a real-time basis in order to provide the dynamic capability required for strategic/tactical training as well as analysis and research.

In the final analysis, it is link connectivities that much be modeled in order to successfully describe EW and C^3 system effectiveness. Link connectivity can be best described with an example. A transmitter to receiver is one link (T_1-R_1) a jammer to the same receiver is another (J_1-R_1). Comparing the two links, the one with the highest signal-to-noise ratio at the receiver (R_1) is the predominant link. When the signal-to-noise ratio of the jammer reaches a certain predetermined value of the signal-to-noise ration of the transmitter at the receiver, then the jamming is judged to commence being effective. What in effect is happening is that one link connectivity (T_1-R_1) is being overridden by a second link (J_1-R_1). A sliding scale of degradation and effectiveness based on receiver/transmitter/jammer locational geometry, systems' capabilities and environmental factors would need to be used to most accurately portray EW interactions. The same process, matching link connectivity, can be used to judge the effectiveness of jamming systems, intercept systems, DF systems and deception systems. I.e., if the signal to noise ration of a specific link, be it T_1-R_1, J_1-R_1, or T_{25}-R_1, is beyond a certain, predetermined value, then the link is effective.

Game Play:

The following is a brief description of the procedures envisioned to be required to adequately war game EW/C^3CM. Game play would be initiated by a player (Team A) attempting a specific tactical action, such as the intercept of an aur raid or changing the disposition of his fleet. In order for the Naval War Gaming System's computer to understand exactly what is happening the player would need to use a specific key word from a furnished list. The system would then be alerted when the war gamer did some specific action such as hitting the "TRANSMIT" button on the keyboard. (This method of alertment/action initiation presupposes that all orders in the war game are or can be transmitted via CRT I/O devices such as the Interactive Console Station of the Naval War Gaming System.

The use of a key world will allow the action specified to be routed to the computer automatically. The computer must have tables in its memory which would specify which emitters are required and what emitters would be helpful to accomplish the task specified by that key word. Team A has the opportunity of designating whether they were turning on or shutting down any specific emitter that was not on the "NEED TO HAVE" list, be it a piece of communications gear, a radar, a navigation aid, a jammer or a deception device (with a concomitant enhancement or reduction of mission effectiveness). The EMCON (Emission Control) posture of the units will also be considered. The Naval War Gaming System would then search its files to see what systems, active or passive, on Team B's side (the opposition) are within range. This is accomplished by comparing the position of each Team B platform and the range of its systems with the position of Team A's platform and the target emitters. If Team B has an ESM system within range, then notice will be sent to Team B's EW module for evaluation and relay to is command module that a detection had taken place, listing what was detected, e.g., an SPS-48 radar, a TOPSAIL radar, HFTTY, HF link 11, etc. If Team B has a DF system within range then a line-of-bearing is also sent to Team B's EW module for evaluation and relay to its command module. If Team B has a jammer or an applicable deception device within range and Team B has elected to turn them on, then the effectiveness of Team A's emitter(s) would be judged to be degraded by a factor predicated on a computation of the variables that are dialed into the game (weather, effective radiated power, sensitivity) and that are computed at that time (range, propagation). Team A is then notified that the effectiveness of its specific tactical action had been degraded by jamming or deception, as appropriate; however, teams will have to wait until the "hot washup" to find out which of their signals had been passively exploited and what the results were. This procedure melds both real war realities (generally, one does not know which of its signals are being exploited) with the proven training/educational technique of immediate feedback.

Imitative deception signals could be simulated just as the signals they are imitating are simulated. Detection and reporting of those signals would flow through the system just as a real signal (which is what happens in the real world.)

The procedures described above place specific information requirements on the war gaming system. The interconnectives and capabilities of both sides' EW and C^3 systems would need to be described. Information of the interconnectivities and capabilities of the US Navy is available and must be continuously updated via fleet communications operating instructions and operating orders. Soviet interconnectivities and capabilities need to be developed and maintained form information available at the SECRET level, the classification level of most war games. Our knowledge of Soviet capabilities and interconnectivities is the most sensitive aspect of this whole question, but enough information is available at the SECRET level to realistically simulate Soviet EW and C^3 systems and their functioning. Indeed, to model one side and not the other's EW or C^3 would place an unrealistic bias in this entire procedure and skew results to the point that untrue, and possibly unwise, tactical lessons would be inferred, one of the primary problems we are seeking to overcome by introducing more realistic EW/C^3CM play into war games.

The envisioned EW/C^3CM subsystem need not be designed for fine-grain analysis of EW and C^3 systems. To do so would make the system interaction calculations significantly more difficult, requiring much more computer power, and yet not contribute significantly to the fleet operator's understanding of the tactical value and dangers of EW/C^3CM. Required figures of merit for system versus system calculations could be derived from studies done by analytical systems such as the ones at the Applied Physics Laboratory and the Naval Ocean Systems Center as well as the Naval Postgraduate School.[13]

As envisioned by this proposal, and that is what it is at this time, the EW/C^3CM aspect of war gaming could be played either as an end in itself–as an EW/C^3CM war game –or be integrated into regular war games, with only the results impinging on the interaction of the opposing forces. The use of the effective radiated power, sensitivity and frequency range in user-controlled tables would permit the changing of the emphasis of the game in mid-game. Basically, the EW/C^3CM war game subsystem outlined above requires computer resident indices of equipments, cross-referenced to platforms, tables of equipments' characteristics and capabilities, and matrices of systems required and systems "nice to have" for each specific tactical action as well as a matrix of each system's vulnerabilities, expressed as a figure of merit, in order to calculate and express overall system measures of effectiveness.

The above subsystem would be sufficiently granular to provide was game players with immediate feedback as to their EW/C³CM vulnerabilities and deficiencies as well as their EW/C³CM assets and capabilities. By using proven figures of merit the results would be scientifically verifiable while demonstrating to the players their need to fully consider the electronic "weapon," the "battle for the fourth dimension," in all tactical decisions. The subsystem could be used to develop and test doctrine and tactics as well as provide training and educations.

The choice is clear. The status quo is unsatisfactory and fine-grain analysis is not required to educate and train fleet operators. A system similar to what is outlined in this paper should be implemented at the Naval War College as soon as possible. Clearly more work needs to be done on this concept but this paper is a start from which further definition and refinement can be made. In the end, hopefully, the Navy will have a capability to educate and train its commanders to think and fight in terms of the "Fourth Dimension."

Notes

1. Speech by RAdm. A.A Gallotta, Director, Electronics Warfare and Cryptology, Office of the Chief of Naval Operations, at the Naval War College, 21 May 1982.
2. Robert B. Batburst, *Understanding the Soviet Navy: A Handbook* (Newport, R.I.: Naval War College Press, 1979), p. 127
3. Defense Science Board, Task Force on Navy Counter $C^3(U)$. (Washington, D.C.: April 1979), p.19, SECRET
4. M.P. Arrazhev, V.A. H'uin and N.P. Mar'yin, *Electronic Warfare – Borba s radioelectronnynii seedstvanni* (Mosvow: Voyenizdat, 1972), translated by Leo Kanner Associates, Redwood City, Calif,. p.2-7. V. Grankin, *Electronic Warfare*— (Moscow, Voyennyy Vestnik, Nr. 4, April 1977), translated by Translation Division Foreign Technology Division, WP-AFB, Ohio.
5. John M. Collins, U.S. – *Soviet Military Balance 1960-1980* (New York: McGraw-Hill, 1980) p.233.
6. Walter P. Senio, Systems Research Laboratory, *A Study of Soviet C^3, Electronics Warfare, and SIGINT Technology* (U), (Washington, D.C.. Defense Intelligence Agency, March 1982), p.I.I. SECRET.
7. US Dept. of Defense, *Command, Control and Communications Contermeasures (C^3CM)*, DODINST 4600.4 (Washington: 1979), p.1.
8. A.A Gallotta, "Training-along the Road to Combat Effectiveness," *Journal of Electronic Reference,* September/October 1981, p.22
9. For a more complete unclassified description of Soviet REC capabilities see Guy Thomas, "Soviet Radio Electronic Combat and the U.S. Navy," *Naval War College Review,* July-August 1982, p.16.
10. RAdm. A.A. Gallota speech.
11. Interview with Dr. Frank Shoup, Deputy Director for Studies and Analysis, Systems Analysis Division, Office of the Chief of Naval Operations, 8 March 1982.
12. Robert Brandenburg of Naval Ocean Science Center, San Diego, Calif., Thomas P. Roma of Boeing Aerospace Company, Seattle, Wash., Charles A. Gettier of Illinois Institute of Technology Research Institute, Annapolis, Md., and Cdr. Gallagher of the Data Link Vulnerability Joint Test Force, Kirtland AFB, NM., have each written papers that have substantial input to the formulization of this study's concepts of how EW/C^3CM could be incorporated into war gaming at the Naval War College.

13. *The Applied Physics Laboratory of the Johns Hopkins University* concluded a study using their Battle Analyzer was gaming facility to determine the level of antijamming protection a naval task force would require in nine scripted scenario war game events. In order to perform their analysis, the Applied Physics Laboratory (and several other organizations) have created statistical descriptions of HF and UHF propagation (a central necessity for the accurate war gaming of EW and C^3 systems).

LCdr. Thomas is a member of the Center for War Gaming Staff at the Naval War College.

Made in the USA
Middletown, DE
15 May 2024

54242636R00203